有机化学实验

Experimental Organic Chemistry

for Students of Medicine and Biology

（双语）

主　编　冯文芳

副主编　陈东红　叶晓霞

编　者　陈东红　华中科技大学

Huazhong University of Science and Technology

陈国辉　中南大学

Central South University

冯文芳　华中科技大学

Huazhong University of Science and Technology

付世涛　华中科技大学

Huazhong University of Science and Technology

王　艰　福建医科大学

Fujian Medical University

许秀枝　福建医科大学

Fujian Medical University

叶晓霞　温州医学院

Wenzhou Medical College

于姝燕　内蒙古医科大学

Inner Mongolia Medical University

华中科技大学出版社

中国·武汉

内容提要

本书是主要供生物、环境和医药类专业学生短学期使用的基础有机化学实验双语教材。全书包括四个部分：有机化学实验的基本知识、基础实验技能训练、有机化合物的基本鉴定以及有机化合物的合成综合训练。其中基础实验技能训练有12个实验，有机化合物的基本鉴定有4个实验，有机化合物的合成综合训练有5个实验。本书的特色是内容精练，每一部分内容都有相对应的完整的英文，每一个实验都有详细的预习和操作实践指导，以便于教学使用。

本书可作为高等院校非化学专业的本科生和留学生的教材，也可供从事双语和全英语教学的相关人员参考。

图书在版编目(CIP)数据

有机化学实验(双语)/冯文芳　主编. —武汉：华中科技大学出版社，2013.12(2024.1重印)
ISBN 978-7-5609-9563-2

Ⅰ.①有…　Ⅱ.①冯…　Ⅲ.①有机化学-化学实验-教材-汉、英　Ⅳ.①O62-33

中国版本图书馆CIP数据核字(2013)第299874号

有机化学实验(双语)　　冯文芳　主编

策划编辑：王新华
责任编辑：王新华
封面设计：李　嫚
责任校对：张　琳
责任监印：周治超
出版发行：华中科技大学出版社(中国·武汉)　　电话：(027)81321913
　　　　　武汉市东湖新技术开发区华工科技园　　邮编：430223
录　　排：华中科技大学惠友文印中心
印　　刷：武汉邮科印务有限公司
开　　本：787mm×1092mm　1/16
印　　张：12.75
字　　数：328千字
版　　次：2024年1月第1版第4次印刷
定　　价：28.00元

前　言

全球化和科技进步正在越来越迅速地改变着当今大学的办学模式。为了实现最大程度的资源共享，很多高等院校都进行了新一轮合并和重组，从而使专业公共基础课所面对的教学对象更加庞大和多元化。面对各种教学需求，有针对性地进一步细分教材是非常有必要的。比如就目前华中科技大学化学与化工学院的教学情况来看，有机化学实验课程教学面临相当一部分生物、环境、临床医学、预防医学、护理、口腔、药学等专业学生的短学期学习需求，其中有部分专业还需要采用双语或全英语教学，此外还有每年的临床专业留学生的全英语教学。这些学生在短学期内需要了解和掌握有机化学实验的基本知识，具备基本实验操作技能，以便为后续实验课程打下一定的基础。其他综合性和医科类大学的有机化学实验教学也面临着相似的情况。本书正是为了满足新时期的教学需求，主要供生物、环境和医药类专业学生短学期使用的基础有机化学实验双语教材。

本书的主要特点如下：

(1) 突出基础有机化学实验的通识内容。包括有机化学实验的基本知识，有机化学实验的基本实验技能，有机化合物的基本鉴定，以及有机化合物的合成综合训练。其中以基本实验技能为教学重点。

(2) 实验内容全面而精练，并全部配以相应的英文。英文实验紧随中文之后，便于读者中英文对照查阅。

(3) 突出实验预习的指导性。每个实验都有预习要求和实验操作指导，便于学生做好充分的预习，具有很强的指导性。

(4) 突出实际操作的便利性。每个实验都列出了所涉及试剂的物理性质与物理常数、实验原理及实验装置图等实验相关的信息，具有很强的实用性。

此外，本书在内容选编上特别加入了一些适合生物及医学等专业学生的实验，比如纸色谱、纸上电泳，以及有机化合物的性质实验等，这在现有的基础有机化学实验教材中是比较少见的。部分实验还安排了多种实验方案供选择使用。另外，在附录中列出了部分常用元素的相对原子质量、常见共沸混合物的性质等，以便查阅。

本书的编写分工如下：华中科技大学冯文芳负责全书的编排与全书的修订和统稿，并编写实验十三至十六和附录；华中科技大学陈东红负责全书中文审稿和修改，并编写实验五、九、十二；华中科技大学付世涛负责部分修改工作，并编写实验一至四；温州医学院叶晓霞负责编写第一部分的第一、二、三节；内蒙古医科大学于姝燕负责编写第一部分的第四、五、六节；中南大学陈国辉负责编写实验六、十、十一；福建医科大学许秀枝负责编写实验七、十七至十九；福建医科大学王艰负责编写实验八、二十、二十一。

感谢华中科技大学出版社的大力支持以及有关工作人员的辛勤付出。感谢各参编院校有关领导的支持和鼓励。

限于编者水平，书中不妥之处在所难免，恳请各位读者批评指正。

编　者

2013 年 10 月

目　录

第一部分　有机化学实验的基本知识

第一节　有机化学实验的基本规则

有机化学是一门以实验为基础的学科，学习有机化学必须认真做好有机化学实验。通过实验可得到基本实验技能的全面训练，同时也能验证、巩固和加深课堂讲授的基本理论和基本知识，培养观察能力、分析问题和解决问题的能力，实事求是的科学态度，严谨细致的科学作风和良好的实验工作习惯，为以后进一步的学习和工作打下扎实的基础。

为了保证实验的顺利进行和实验室的安全，所有学生在进入有机化学实验室前都必须遵守以下相关规定：

(1) 实验前学生要认真准备，明确实验的目的、要求和内容，了解实验的基本原理、方法和步骤。必要时，学生应该提前写出实验设计或草案。

(2) 熟悉实验室工作的安全规则，学习如何正确使用水、电、煤气、防毒罩、灭火器以及实验仪器。

(3) 进入实验室时要穿实验制服，严禁穿拖鞋入内。严格遵守实验室安全规程。如果有任何事故发生，应及时报告教师。

(4) 实验前清点实验仪器，如果发现有破损，应立即向教师报告，按规定在实验准备室换取。

(5) 在使用化学品时，应仔细阅读标签，只使用实验需要的药品。用后立即塞上塞子，避免塞子混乱以及化学品污染。使用固体试剂时，必须保持药匙清洁、干燥。

(6) 在实验室里，学生应该保持安静，遵守纪律。未经许可，不得离开实验室。在实验过程中，认真操作，仔细观察，积极思考，如实记录。

(7) 在实验室，各种固体或液体废物应该严格按照要求分类回收处理。

(8) 离开实验室之前，仔细检查水、电和煤气是否已关掉，并用肥皂和水彻底洗手。

第二节　有机化学实验的安全知识

有机化学品因其易燃性、易爆性、毒性、腐蚀性和挥发性，而属于危险品。同时，对于常用的玻璃仪器及电器、设备，如果操作不正确，也有可能导致实验事故。因此，有机化学实验室对于学生来讲是一个具有潜在危险性的地点，进入实验室工作的人员需要具有高度的安全防范意识。

一、防火

(1) 处理易燃溶剂(如苯、乙醚、丙酮、石油醚、二硫化碳和乙醇)时，应远离明火。

(2) 当回流或蒸馏液体时，应该把沸石放入烧瓶中，防止液体因过热而冲出。

（3）使用酒精灯时，为了避免酒精外溢，引起火灾，严禁使用酒精灯点燃其他酒精灯。

（4）经常检查煤气管阀、煤气灯是否完好，以防止漏气。

一旦发生着火事故，应沉着镇静并及时处理，一般采用如下措施：

（1）防止火势扩散：立即熄灭附近火源，切断电源，移开尚未着火的易燃物。

（2）根据火势立即灭火：若火势较小，可用石棉布、黄沙盖熄；如着火面积大，立即取灭火器灭火。有机物着火时，勿用水浇；电器着火时，应切断电源，然后用干粉灭火器灭火。

二、防爆

（1）安装常压蒸馏装置时，避免系统密闭。由于瓶中干渣可能存在过氧化物或其他易燃物质，蒸馏时切忌蒸干。

（2）一些有机化合物接触氧化剂时，可能产生剧烈的爆炸或燃烧，要小心搬运和储存。

（3）对反应过于剧烈的实验，应严格控制加料速度和反应温度，使反应缓慢地进行。

三、防毒

（1）严禁在实验室吃、喝或品尝任何试剂。

（2）使用有毒试剂时必须戴橡胶手套，防止接触擦伤的皮肤。不要将有毒试剂倒入下水道。操作后立即洗手。如果操作毒性实验，应在通风橱内完成。

（3）在闻试剂的气味时，不要用鼻子直接去闻。

（4）使用通风橱时，不要把头伸入通风橱内。

（5）实验过程中产生的有毒残渣必须妥善且有效地处理，不准乱丢。

（6）如果出现中毒症状，应立即就医。如毒物已溅入口中，尚未咽下的立即吐出，用大量水冲洗口腔；如已吞下，应根据毒物的性质先进行如下处理：

① 吞下酸：先饮大量水，然后服用氢氧化铝膏、鸡蛋白、牛奶，不要吃呕吐剂。

② 吞下碱：先饮大量水，然后服用醋、酸果汁、鸡蛋白、牛奶，不要吃呕吐剂。

③ 吞下刺激性及神经性毒物：先服用牛奶或鸡蛋白将之冲淡缓和，再将一大匙硫酸镁（约30 g）溶于一杯水中饮下催吐。有时也可用手指伸入喉部促使呕吐。

④ 吸入气体中毒：将中毒者迅速搬到室外，解开衣领及纽扣，若是吸入氯气或溴气，可用稀 $NaHCO_3$ 溶液漱口。

四、防化学灼伤

处理放热性和腐蚀性化学品时应该非常小心，不要让其直接接触身体的任何部分，以免被灼伤。灼伤后立即用大量的自来水洗灼伤部位。

（1）酸灼伤：眼睛灼伤用1% $NaHCO_3$ 溶液清洗；皮肤灼伤用5% $NaHCO_3$ 溶液清洗。

（2）碱灼伤：眼睛灼伤用1%硼酸溶液清洗；皮肤灼伤用1%～2%乙酸溶液清洗。

（3）溴灼伤：立即用酒精清洗，再涂上甘油，或敷上烫伤油膏。灼伤严重者经急救后速去医院治疗。

第三节　废物的处理

实验操作中会产生不同种类的固体或液体废物。废物处理一直是现代社会关注的主要环境问题。实验室废物的处理可以按照下列方式进行：

（1）所有实验室产生的垃圾应分类清楚，并且得到妥善处理。一些难以处理的危险废物应该送到环保部门进行特殊处理。

（2）少量的酸或碱应先中和，然后用大量的水稀释，再冲入下水道。

（3）有机溶剂应储存在通风良好的地方再被回收。

（4）一些致癌物质和疑似致癌物质必须小心回收，避免接触身体。

（5）对与水发生剧烈反应的化学品，处置之前要用适当的方法在通风橱内进行分解。

（6）对于无害的固体废物，可直接倒入普通的废箱中，不应与其他有害固体相混；对有害固体废物，应放入带有标签的广口瓶中。

第四节 常用玻璃仪器和实验装置

一、常用玻璃仪器

1. 普通玻璃仪器

有机实验室常用的玻璃仪器如图 1-1 所示。

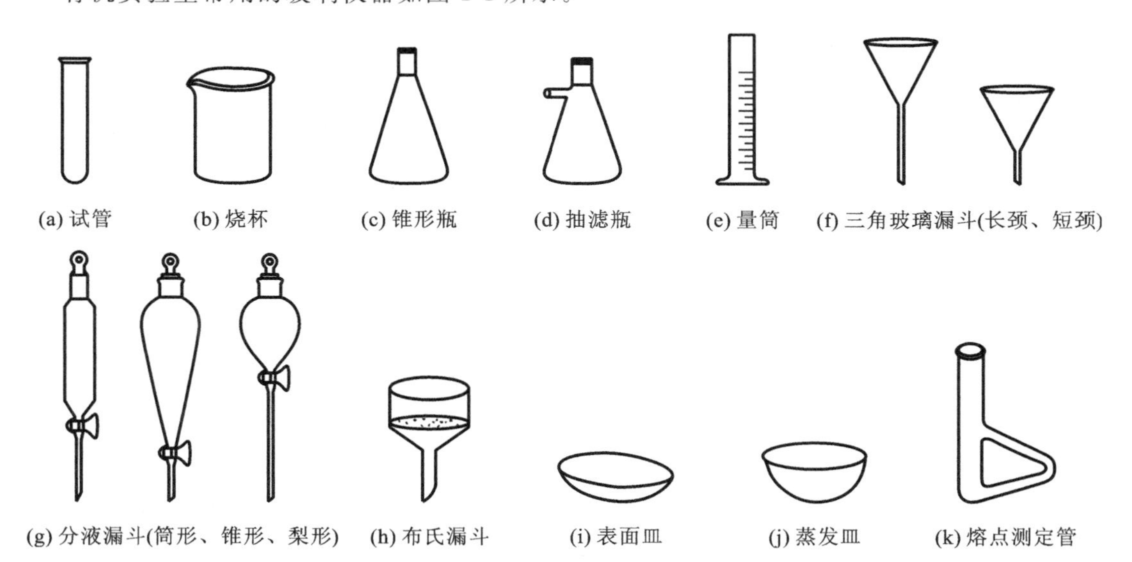

图 1-1 有机化学实验常用普通玻璃仪器

2. 标准磨口玻璃仪器

常见标准磨口玻璃仪器如图 1-2 所示。

3. 常用玻璃仪器的用途

各种常用玻璃仪器的用途如下：

（1）圆底烧瓶：能耐热和承受反应物（或溶剂）沸腾以后所发生的冲击震动。在有机化合物的合成和蒸馏实验中最常使用，也常用作减压蒸馏的接收器。

（2）梨形烧瓶：性能和用途与圆底烧瓶相似。它的特点是在合成少量有机化合物时在烧瓶内保持较高的液面，蒸馏时残留在烧瓶中的液体少。

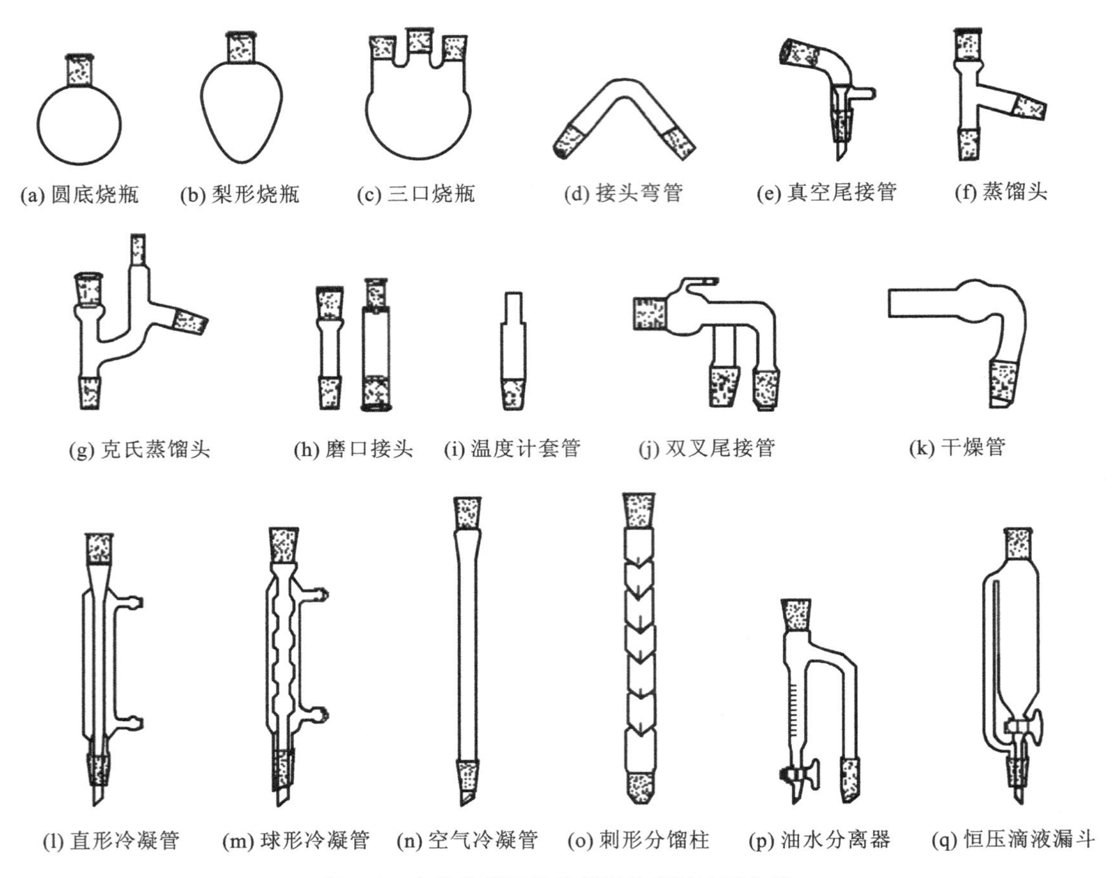

图 1-2　有机化学实验常用标准磨口玻璃仪器

(3) 三口烧瓶:最常用于需要进行搅拌的实验中。中间瓶口装搅拌器,两个侧口装回流冷凝管和滴液漏斗或温度计等。

(4) 锥形烧瓶(简称锥形瓶):常用于有机溶剂进行重结晶的操作,或有固体产物生成的合成实验中,因为生成的固体物容易从锥形瓶中取出来。通常也用作常压蒸馏实验的接收器,但不能用作减压蒸馏实验的接收器。

(5) 直形冷凝管:蒸馏物质的沸点在 140 ℃以下时,要在夹套内通水冷却;超过 140 ℃时,冷凝管往往会在内管和外管的接合处炸裂,故不宜使用。微量合成实验中,可用于加热回流装置上。

(6) 空气冷凝管:当蒸馏物质的沸点高于 140 ℃时,常用它代替通冷却水的直形冷凝管。

(7) 球形冷凝管:其内管的冷却面积较大,对蒸气有较好的冷凝效果,使用于加热回流的实验。

(8) 刺形分馏柱:又称韦氏分馏柱,是每隔一段距离就有一组向下倾斜的刺状物,且各组刺状物间呈螺旋状排列的分馏管。蒸气在分馏柱内的上升过程中,类似于经过反复多次的简单蒸馏,使蒸气中低沸点的成分含量逐步提高。适合于分离少量几种沸点相近的混合物的液体。

(9) 分液漏斗:用于液体的萃取、洗涤和分离,有时也可用于滴加试剂。

(10) 滴液漏斗:能把液体一滴一滴地加入反应器中,即使漏斗的下端浸没在液面下,也能够明显地看到滴加的快慢。

(11) 恒压滴液漏斗:用于合成反应实验的液体加料操作,也可用于简单的连续萃取操作。

(12) 布氏漏斗:为瓷质的多孔板漏斗,在减压过滤时使用。小型玻璃多孔板漏斗用于减压过滤少量物质。

(13) 油水分离器:根据水和有机物的密度差,利用重力沉降原理去除水分的分离器。常用于有水生成的反应体系装置中,用以不断除去反应过程中产生的水。

二、常用实验装置

有机化学实验室常用的实验装置如下。

1. 回流装置

回流装置如图 1-3、图 1-4、图 1-5 所示。

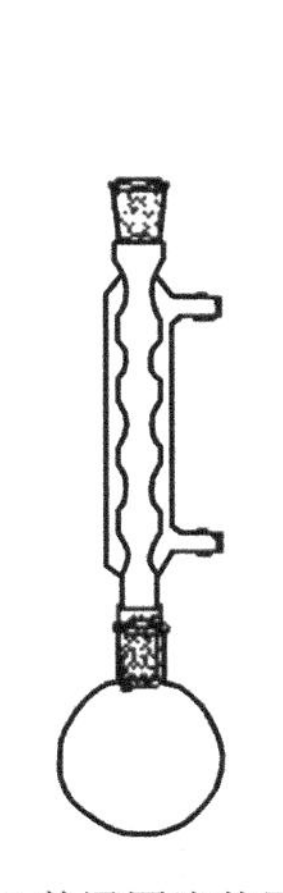

(a) 普通回流装置

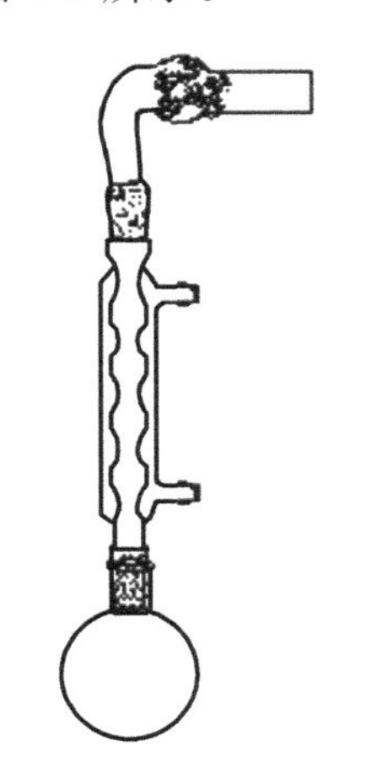

(b) 带干燥管的回流装置

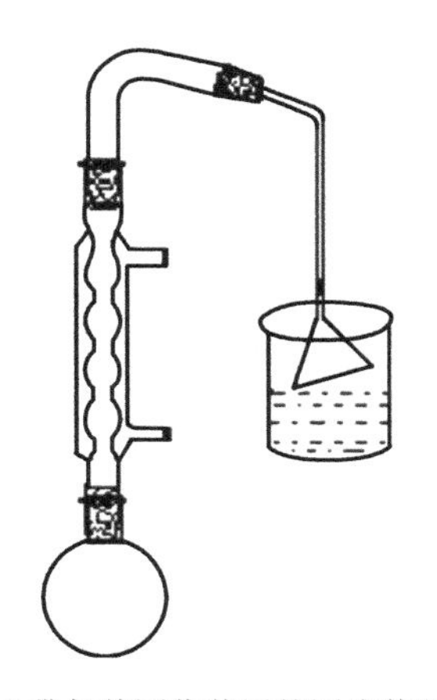

(c) 带气体回收装置的回流装置

图 1-3　回流冷凝装置

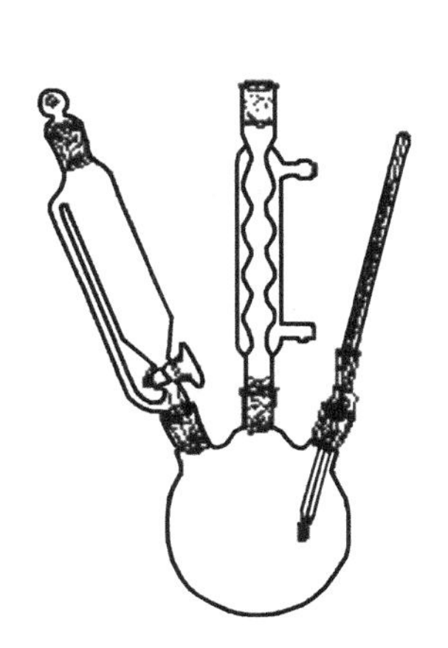

图 1-4　滴加回流装置

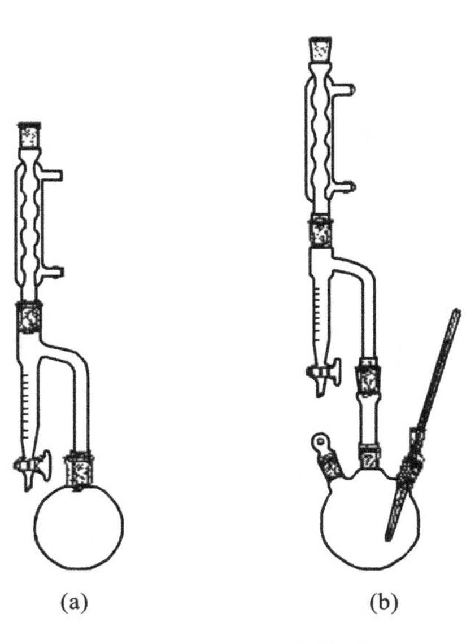

图 1-5　回流分水装置

2. 蒸馏装置

常压蒸馏装置如图 1-6 所示。减压蒸馏装置如图 1-7 所示。

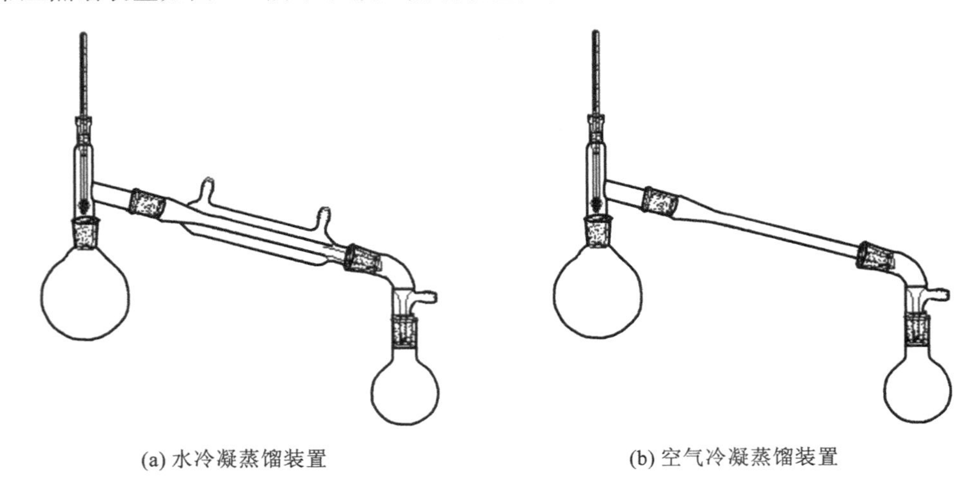

图 1-6　常压蒸馏装置

3. 分馏装置

分馏装置如图 1-8 所示。

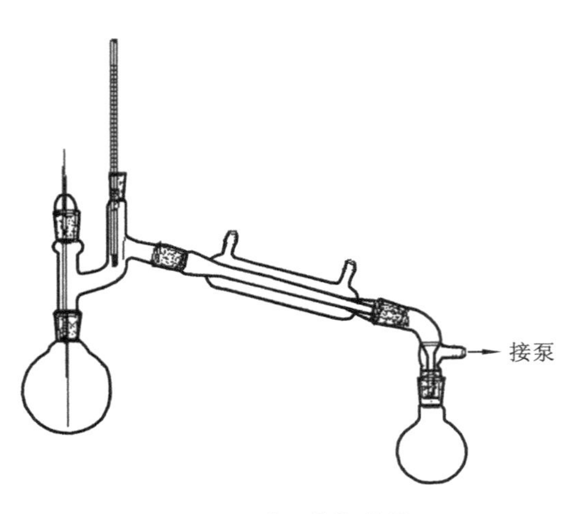

图 1-7　减压蒸馏装置

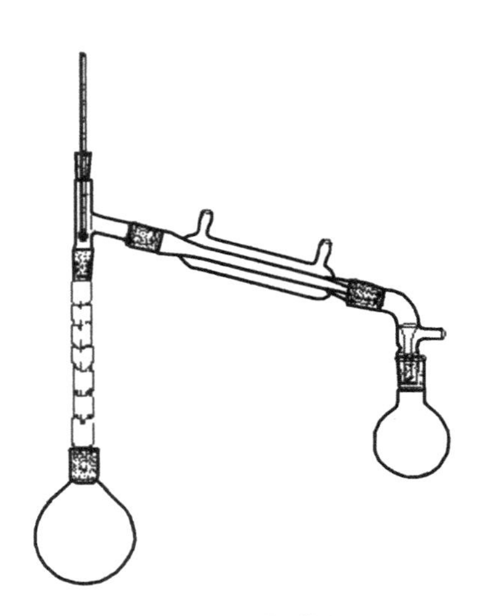

图 1-8　分馏装置

三、使用玻璃仪器的注意事项

使用玻璃仪器的注意事项如下：

(1) 使用时要轻拿轻放，以免弄碎；

(2) 除烧杯、烧瓶和试管可使用正确方法加热外，其他玻璃仪器均不能用火直接加热；

(3) 锥形瓶、平底烧瓶不耐压，不能用于减压操作；

(4) 带活塞的玻璃器皿用后应及时清洗，并在活塞与磨口之间垫上纸片，以防粘连；

(5) 温度计的水银球玻璃很薄、易碎，使用时应小心。不能将温度计当作搅拌棒使用；温度计使用后应先自然冷却后再冲洗，以免破裂；测量温度不得超出温度计刻度范围。

四、安装实验装置的注意事项

安装实验装置的注意事项如下：

(1) 所用玻璃仪器和配件要干净，大小要合适；

(2) 电热套应始终放置在升降台上，搭建装置前应将升降台升至一定高度后，首先按照电热套的高度安装、固定反应瓶，然后顺次连接其他玻璃仪器；

(3) 搭建实验装置时，应按照“由下至上、由左至右”原则，逐个装配，铁夹尽可能夹在磨口接头处且不可夹得过紧；

(4) 反应回流装置应在回流停止后再开始拆卸，拆卸时应按照“由右至左、由上至下”原则，逐个拆除；

(5) 常压下进行的反应，实验装置应与大气相通，保证气密性但不能密闭；

(6) 实验装置要求做到严密、正确、整齐、稳妥。磨口连接处应无缝隙。

五、玻璃仪器的清洗

玻璃仪器用毕后应立即清洗，一般的清洗方法是将玻璃仪器和毛刷淋湿，用毛刷蘸取肥皂粉或洗涤剂，洗刷玻璃器皿的内外壁，除去污物后用水冲洗；当洁净度要求较高时，可依次用洗涤剂、蒸馏水(或去离子水)清洗；也可用超声波振荡仪来清洗。

严禁盲目使用各种化学试剂或有机溶剂来清洗玻璃器皿。这样不仅造成浪费，而且可能带来危害，对环境造成污染。

六、玻璃仪器的干燥

干燥玻璃仪器的方法通常有以下几种：

(1) 自然干燥：将仪器倒置，让水自然流下，晾干。

(2) 烘干：将仪器放入烘箱内烘干，仪器口朝上；也可用气流干燥器烘干或用电吹风吹干。

(3) 有机溶剂干燥：急用时可用有机溶剂助干，用少量95%乙醇或丙酮荡涤，把溶剂倒回至回收瓶中，然后用电吹风吹干。

第五节　常用有机溶剂的性质及使用方法

一、乙醚($C_2H_5OC_2H_5$)

M_w 74.12，b.p.34.6℃，d_4^{20} 0.7134，n_D^{20} 1.351～1.353。

15℃时乙醚可吸收1.2%的水，20℃时乙醚在水中的溶解度约为6.5%，与水共沸物含水1.26%，34.15℃沸腾。乙醚能和绝大多数有机溶剂任意混合。

乙醚沸点低，易挥发、易燃，使用乙醚时严禁明火。在空气中和光照作用下，乙醚极易产生爆炸性的过氧化物，因此乙醚应储存于氢氧化钾中，它能直接将产生的过氧化物转化成不溶性的盐，同时也是一种合适的干燥剂。在使用乙醚前要检测过氧化物，取少量乙醚，加等体积的2%碘化钾水溶液和几滴稀硫酸，振摇，再加1滴淀粉试液，呈紫蓝色即表示有过氧化物存在。用酸性硫酸亚铁溶液洗涤乙醚可除去过氧化物。蒸馏久置的乙醚时切忌蒸干，以免因过氧化物产生爆炸。

乙醚蒸气有麻醉作用,使用时应戴防护口罩,并保持室内通风。

二、乙醇(C_2H_5OH)

M_w 46.07,b.p.78.5℃,d_4^{20} 0.789,n_D^{20} 1.3614。

乙醇为无色透明液体,易燃,能以任意比例与水、乙醚、氯仿和苯混合。对人体的毒性较低。能与水形成共沸物,沸点为78.17℃,含乙醇96%。许多极性和弱极性的有机化合物能溶解在乙醇中,因此乙醇是重结晶有机化合物的良好溶剂。市售乙醇的含量为95%,根据对无水乙醇纯度的要求不同可选择不同的纯化方法。

1. 无水乙醇(含量为99.5%)的制备

在250 mL圆底烧瓶中,放入45 g生石灰、100 mL乙醇(95%),装上回流冷凝器(上接一个无水氯化钙干燥管),在水浴上回流2～3 h,然后改为蒸馏装置蒸馏,弃去少量前馏分后收集得无水乙醇。

2. 绝对乙醇(含量为99.95%)的制备

在250 mL圆底烧瓶中,将2 g金属钠加入100 mL纯度至少是99%的乙醇中,加几粒沸石,装上球形冷凝器(上接一个无水氯化钙干燥管),回流30 min。再改成蒸馏装置蒸馏,收集得绝对乙醇。若要制备纯度更高的绝对乙醇,则可在回流30 min后,加入4 g邻苯二甲酸二乙酯(金属钠虽能与乙醇中的水作用,产生氢气和氢氧化钠,但所生成的氢氧化钠又与乙醇发生反应,所以单独使用金属钠不能完全除去乙醇中的水,须加入过量的高沸点酯,如草酸二乙酯与生成的氢氧化钠作用,抑制上述反应,从而达到进一步脱水的目的),再回流10 min,然后改成蒸馏装置蒸馏,收集产品即得。因为乙醇具有非常强的吸湿性,所以在操作时,动作要快,尽量减少转移次数以防止空气中的水分进入,同时所用仪器必须事前干燥好。

含水超过0.05%的乙醇与三乙氧基铝的苯溶液产生大量白色沉淀。

三、丙酮(CH_3COCH_3)

M_w 58.08,b.p.56.5 ℃,d_{25}^{25} 0.788,n_D^{20} 1.3591。

丙酮易燃易挥发,能与水、乙醇、乙醚以任意比例互溶,不能与水形成共沸物。丙酮是很多有机物质的良好溶剂。普通丙酮常含有少量的水及甲醇、乙醛等还原性杂质。市售丙酮的纯度可满足大多数要求。高纯度丙酮可用五氧化二磷干燥,注意用碱性干燥剂时会得到缩合产物。

四、苯(C_6H_6)

M_w 78.11,b.p.80.1 ℃,d_4^{15} 0.8786,n_D^{20} 1.5011。

苯是无色透明的液体,易燃。难溶于水,20℃时苯可吸收0.06%的水,同温下苯在水中的溶解度约为0.07%。69.25℃与水的共沸物含水8.83%,常利用苯与水共沸的性质来除去反应中生成的水。苯是非极性溶剂,常用于提取和重结晶有机化合物。

苯具有很强的血液毒性,能通过皮肤吸收,长期接触会引起慢性中毒,主要表现为破坏人体造血功能。

五、乙酸乙酯($CH_3COOC_2H_5$)

M_w 88.11，b.p.77 ℃，d_4^{20} 0.902，n_D^{20} 1.3720。

乙酸乙酯为无色易燃液体，能与多数有机溶剂混合，100 mL 水中能溶解 8.6 g 乙酸乙酯。乙酸乙酯与水的共沸物沸点为 70.38 ℃。乙酸乙酯是许多有机化合物的良好溶剂，但它能与胺类起反应，精制胺类化合物时不能用乙酸乙酯做溶剂。

市售乙酸乙酯一般含有少量水、乙醇和乙酸。纯化时可用等体积的 5%的碳酸钠溶液洗涤，再用氯化钙干燥，蒸馏。更高要求的纯化可用五氧化二磷干燥，过滤，然后在干燥条件下蒸馏。

六、三氯甲烷($CHCl_3$)

M_w 119.38，b.p.61～62 ℃，d_4^{20} 1.484，n_D^{20} 1.4476。

三氯甲烷又称为氯仿，是无色透明液体，微溶于水，蒸气不燃烧。三氯甲烷能溶解许多有机化合物，实验中可用它萃取和精制有机化合物。三氯甲烷能与水形成共沸物，沸点为 61 ℃。

三氯甲烷在日光下易氧化成氯气、氯化氢和光气(剧毒)，故应储存于棕色瓶中。市售三氯甲烷常含 1%乙醇做稳定剂，以结合分解产生的光气。三氯甲烷不能用金属钠干燥，否则会发生爆炸。三氯甲烷也不能与胺类等碱性试剂接触，因碱能使其分解为二氯卡宾。

三氯甲烷蒸气有麻醉作用，使用时应戴防护口罩，并保持室内通风。长期大量接触会引起肝肾损伤和心律不齐。

七、正己烷 (C_6H_{14})

M_w 86.18，b.p.69 ℃，d_4^{20} 0.659，n_D^{20} 1.3748。

正己烷为无色透明液体，易燃，不溶于水。常用正己烷来提取、精制和层析有机化合物。正己烷中的主要杂质是烯烃和芳香族化合物，可用发烟浓硫酸少量多次萃取，直到酸层为淡黄色，再依次用浓硫酸、水、2%氢氧化钠溶液、水洗涤，然后用氢氧化钾干燥后蒸馏。

八、石油醚

石油醚为轻质石油产品，是低相对分子质量烷烃类的混合物。常用石油醚来提取和层析有机化合物。市售有 30～60 ℃、60～90 ℃、90～120 ℃等沸程规格的石油醚，常含有少量不饱和烃，沸点与烷烃相近，用蒸馏法无法分离。纯化石油醚时通常用等体积的浓硫酸洗涤2～3次，再用 10%硫酸加入高锰酸钾配成的饱和溶液洗涤，直至水层中的紫色不再消失为止。然后用水洗，经无水氯化钙干燥后蒸馏。若需绝对干燥的石油醚，可加入钠丝。

九、四氢呋喃(C_4H_8O)

M_w 72.11，b.p.66 ℃，d_4^{20} 0.8892，n_D^{20} 1.4070。

四氢呋喃(THF)为无色液体，可燃，具有醚类的气味，可溶于水和其他有机溶剂，是有机反应的良好溶剂。63.2℃与水形成共沸物，含四氢呋喃 94.6%。市售四氢呋喃常含水，久贮后可能含有过氧化物(过氧化物的检测和处理同乙醚)，在加碱处理或蒸馏近干时会引起爆炸。

十、N,N-二甲基甲酰胺(C_3H_7NO)

M_w 73.09,b.p.153 ℃,d_4^{25} 0.9445,n_D^{25} 1.4290。

N,N-二甲基甲酰胺(DMF)是可燃性液体,能以任意比例溶于水和大多数有机溶剂。它是极性非质子性溶剂,对许多有机物和盐的溶解度较大,但由于沸点较高,很少用于有机物质的精制。DMF中常含有少量水、氨、胺和甲醛,在常压蒸馏时会部分分解,若有酸或碱存在,分解加快。因此最好用硫酸钙、硫酸镁、氧化钡、硅胶或分子筛干燥后减压蒸馏。如含水较多,可加入1/10体积的苯,在常压、80℃以下蒸去水和苯,然后用硫酸镁或氧化钡干燥,再进行减压蒸馏。DMF必须避光保存,否则见光会分解成二甲胺和甲醛。

DMF蒸气有毒,易通过皮肤被吸收。吸入过多的DMF的蒸气后会引起恶心呕吐。

十一、吡啶(C_5H_5N)

M_w 79.10,b.p.115.2 ℃,d_4^{20} 0.9827,n_D^{20} 1.5085~1.5105。

吡啶具有吸湿性,能与水、醇和醚以任意比例混合。94℃与水形成共沸,含吡啶57%。吡啶具有碱性,常用作有机反应的缚酸试剂。分析纯的吡啶含有少量水分,可供一般实验用。如要制得无水吡啶,可将吡啶与氢氧化钾或氢氧化钠一同回流,然后隔绝潮气蒸出备用。保存干燥的吡啶时应将容器口用石蜡封好。

吡啶易引起皮肤湿疹,吸入吡啶蒸气会引起恶心、肠胃痉挛及神经损害。

十二、乙腈(CH_3CN)

M_w 41.05,b.p.81.6 ℃,d_4^{15} 0.7875,n_D^{15} 1.3460。

乙腈为易燃性液体,能与水、乙醇和乙醚以任意比例互溶。与水在26.7℃形成共沸物,其中含乙腈84.1%。乙腈是极性非质子性溶剂,是许多有机物的良好溶剂,常用于有机反应。乙腈可用五氧化二磷一起煮沸反复回流直到无色,然后蒸馏,再分馏精制。

乙腈有毒,通常含有氢氰酸,应特别注意。

第六节　预习报告及实验报告

一、预习报告

每位学生都应准备实验记录本,在每次实验前必须认真预习并写好预习报告,做好充分准备。预习的具体要求如下:

(1) 按照书中对实验提出的预习要求了解相关内容并做记录;

(2) 按照自己的理解写出方法原理或反应式(主要反应式、主要副反应);

(3) 摘录主要试剂及产物的物理常数以及主要试剂的规格和用量(g、mL、mol),列出实验所需玻璃仪器;

(4) 列出实验相关的大体步骤及有关注意事项;

(5) 列表,用以记录有关实验数据。

二、实验记录

实验中不仅要规范操作，仔细观察，积极思考，还应将观察到的实验现象（如加热情况、颜色变化、pH值的改变、沉淀、气泡等）和数据（如反应温度，产物的熔点、沸点等）及时记录下来，不允许写回忆笔记。实验记录应实事求是、简明扼要、字迹整洁、清楚完整。实验完毕后，学生须将原始记录与产品交由教师检查后方能离开实验室。

三、实验报告

实验完成后，整理有关数据和材料，对实验现象进行分析、归纳、总结，按一定格式及时完成实验报告，并总结实践体会和经验教训，对存在的问题提出改进意见或解决方法。实验报告应条理清楚，书写工整，图表清晰，格式符合要求。实验报告基本格式如下：

实验名称＿＿＿＿＿＿＿＿＿＿＿＿

（一）实验目的和要求

（二）反应原理

（三）主要试剂及产物的物理常数

（四）主要试剂用量及规格

（五）仪器装置

（六）操作步骤及现象记录

（七）产率计算

（八）讨论

Part 1 Fundermentals of Organic Experiments

1.1 General Rules for the Organic Chemistry Lab

In order to ensure all experiments going smoothly and safely, all students must abide by the following rules when entering an organic chemistry lab:

(1) Before the experiment, students must prepare carefully, understand the purposes, requirements and the contents of the experiment; master the basic principles of the experiment, the methods and steps. When necessary, students should write the experiment design or the draft in advance.

(2) Familiarize yourself with the safety rules for lab work and learn about how to correctly use water, power, gas, respirators, fire extinguishers and the experimental instruments.

(3) Wearing a lab coat when entering the lab. Don't wear slippers. Strictly abide by the lab safety regulations. If there is any accident occurred, students should promptly report to the teacher.

(4) Counting experimental instruments before experiment. If you find there is any breakage, you should immediately report to the teacher and get a new one at preparation room by the regulation.

(5) Before using chemicals, read their labels carefully. Use them only as required for the experiment. Cover the stopper of the container immediately after use, and avoid the stopper being confused as well as chemicals being contaminated. When transferring solid reagents, you must keep the spoon clean and dry.

(6) In the lab students should keep quiet and observe the changes. Don't leave the lab without permission. During the experiment you should operate seriously, observe carefully, think positively and record faithfully.

(7) During or after the experiment, all kinds of solid or liquid wastes should be placed in various authorized containers under guidance.

(8) Before leaving lab, check carefully whether water, power and gas are switched off safely, and wash your hands thoroughly with soap and water.

1.2 General Lab Safety

Chemicals are hazardous because of their flammable, explosive, volatile, corrosive and toxic properties. Also, there is the possibility of experimental accidents to glass equipments and electrical appliances if operated incorrectly. Therefore, organic chemistry lab is potentially one of the most dangerous locations for students.

Ⅰ Fire-proof

(1) When handling with flammable solvents, such as benzene, ether, acetone, petroleum ether, carbon disulfide or ethanol, you should keep them far away from the fire.

(2) When refluxing or distilling liquid, you should put zeolite (boiling stones) into the flask. Otherwise the liquid will be overheated and then rush out.

(3) When using alcohol lamp, don't use one of alcohol lamps lighting to another, in order to avoid alcohol overflowing and causing fire.

(4) Always check the gas valve and the gas lights and keep them in good condition from leakage.

Keep calm and treat in time in the fire accident, generally by taking the following measures:

(1) To prevent the fire from spreading: Immediately put out fire at vicinity, cut off power supply, and remove the combustible without fire.

(2) Put out the fire immediately according to intensity of a fire: If the fire area is small, use asbestos cloth or sand to cover the fire; if the fire area is big, use the fire extinguishers. To extinguish fire on organic matter, do not use water; to an electrical fire, cut off the power at first, and then use dry-ice fire extinguishers to put out the fire.

Ⅱ Explosion-proof

(1) The apparatus should be assembled correctly. Do not heat a inclosed system. Distillation to dryness is also a dangerous practice because of the possible presence of peroxides or other explosive materials in the dry residues in the flask.

(2) A fierce explosion or combusting can be produced when some organic compounds come into contact with oxidizers. Beware of their handling and storage.

(3) If the reaction is too violent to control, you'd better turn down the reaction temperature to slow down the reaction reactivity.

Ⅲ Poisoning-proof

(1) Since the reagent bottle might be mislabeled, do not eat, drink or taste any reagent in the lab.

(2) Don't smell the reagent directly by nose.

(3) You must put on rubber gloves when handle with toxic reagents and treat with them in a ventilation cabinet. Keep them away from your mouth or skin, and never pour them into the drain. Wash your hands immediately after operation and move to an area where you can breathe fresh air and rest.

(4) When you operate an experiment in the ventilation cabinet, don't stretch your head into the cabinet. It's important to keep your head out of the front panel.

(5) Toxic residues must be handled properly and effectively. No littering always!

(6) If the symptoms of poisoning appear, you should receive medical treatment immediately.

If the toxic chemicals have splashed into the mouth, spit out immediately, and rinse mouth with plenty of water. If they have been already swallowed, the first-aid should be done according to the nature of the poison as follows:

(1) Swallow acid: Drink plenty of water first, and then take aluminum hydroxide paste, egg white or milk. Don't take vomiting agents.

(2) Swallow base: Drink lots of water first, and then take vinegar, sour juice, egg white or milk. Don't take vomiting agents.

(3) Swallow irritating or nerve poisons: The first aid is to promote a vomiting as quickly as possible. Drink milk or egg white to dilute the poisons at first, and then drink a spoonful of magnesium sulfate (about 30 g) dissolved in a glass of water. Sometimes you can promote vomiting by pressing the throat with a finger.

(4) Inhalation of gaseous poisons: Move to safe area quickly, unlocking the collar and buttons. If inhale chlorine or bromine gas, gargle with dilute solution of $NaHCO_3$.

Ⅳ Prevention of Chemical Burns

Dealing with hot objects and corrosive chemicals should be very careful, don't make the direct contact with any parts of the body. In case of such an accident, wash the affected area immediately with copious amounts of running water.

(1) Acid burns: Clean with 1% $NaHCO_3$ solution if eye burns; clean with 5% $NaHCO_3$ solution if skin burns.

(2) Alkali burns: Clean with 1% boric acid solution when eye burns; clean with 1%～2% CH_3COOH when skin burns.

(3) Bromine burns: Immediately clean with alcohol, and then coated with glycerin, or apply the burn ointment. In case of severe burns go to the hospital quickly after first aid.

1.3 Disposal of Lab Wastes

Experimental operations always produce different kinds of solid or liquid wastes. Waste disposal has been one of the major environmental problems of modern society. The useful disposal of lab wastes can be done in the following ways:

(1) All wastes generated in the lab should be classified clearly before being disposed. Some hard-to-handle hazardous wastes should be delivered to specialized departments for special treatment.

(2) Small amounts of acid or base should be neutralized first and diluted with large amounts of water before flushing down the drain.

(3) Organic solvents should be stored at the well-ventilated place before being recycled.

(4) Some carcinogens and substances suspected of causing cancers must be handled with

great care, avoiding contact with your body.

(5) Chemicals which react violently with water should be decomposed appropriately in the ventilating cabinet.

(6) Some harmless solid wastes can be directly poured into the waste bin, but avoiding being confused with other harmful solid. Those hazardous solid wastes should be put into labeled jars.

1.4 Lab Glass Instrument and Apparatus

Ⅰ Common Glass Instrument

1. Common glassware

A typical set of lab glassware for organic experiments is shown in Figure 1-1.

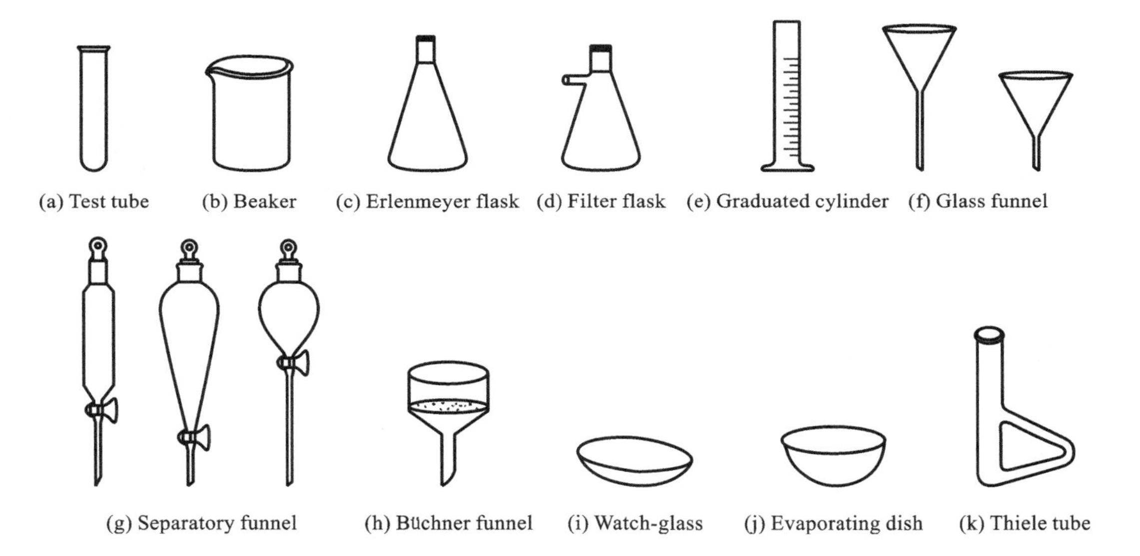

Figure 1-1 Common Glassware for Organic Experiments

2. The standard ground glass instrument

Common standard ground glass equipments are shown in Figure 1-2.

3. Notes

(1) Round-bottom flasks for distillation, reflux;

(2) Three-neck flasks for more complicated reaction set-ups (two-neck flasks are also available);

(3) Erlenmeyer flasks for titration, crystallization, preparation;

(4) Beakers for heating or mixing;

(5) Dropping funnels for adding liquids;

(6) Separatory funnels for extraction and reaction work-up;

(7) Condensers for distillation, relux;

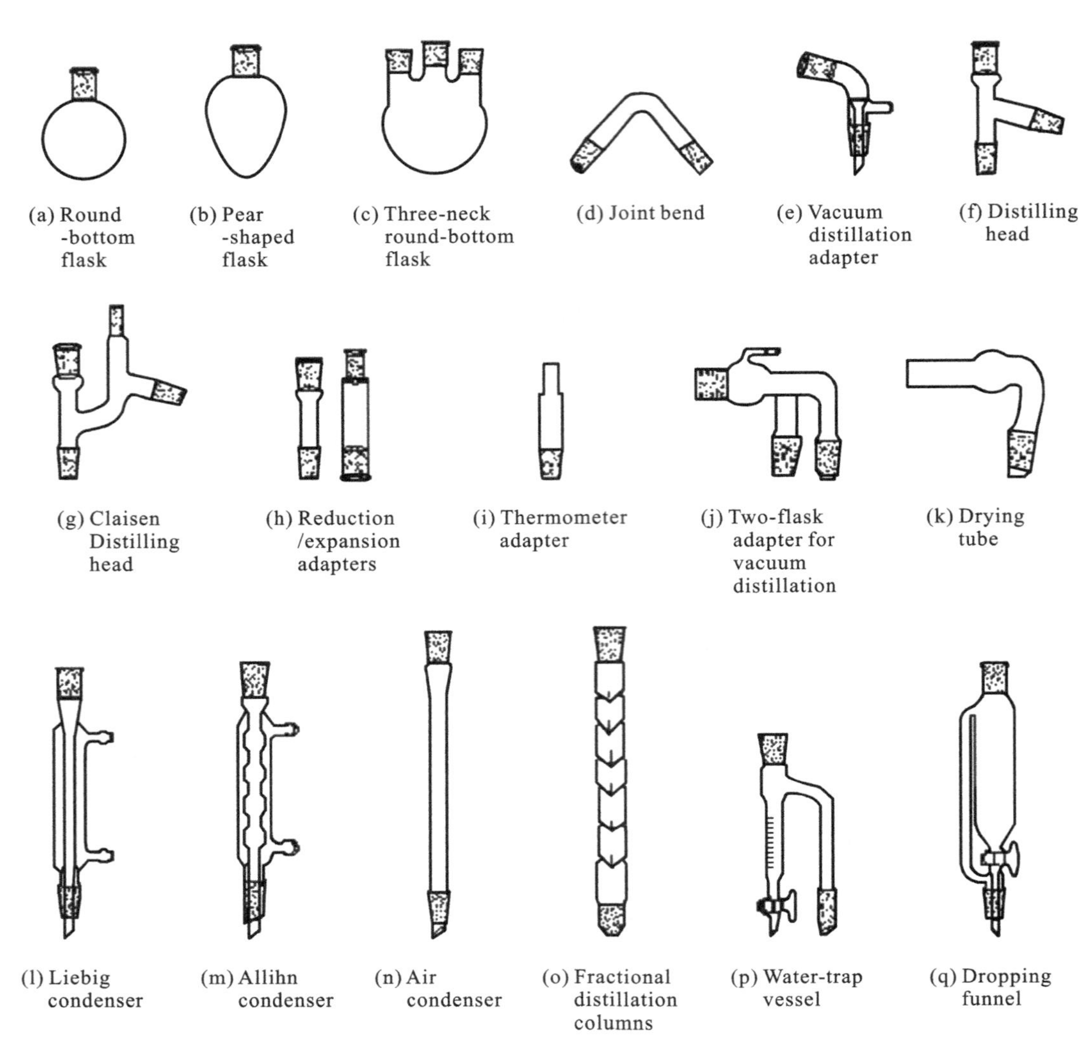

Figure 1-2　Standard Ground Glass Instrument

(8) Air condensers for distillation with high boiling liquids;

(9) Drying tubes for reflux;

(10) Distilling heads for distillation;

(11) Various adapters for distillation, vaccum distillation;

(12) Filter flasks (suction flasks) for collecting the filtrate.

Ⅱ Common Apparatus

1. Reflux apparatus

Common reflux apparatus are shown in Figure 1-3、Figure 1-4 and Figure 1-5.

2. Distillation apparatus

Simple distillation apparatus is shown in Figure 1-6, and vacuum distillation apparatus is shown in Figure 1-7.

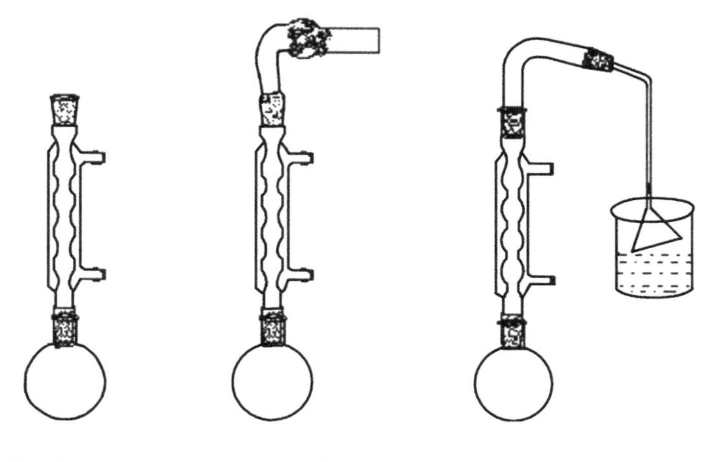

(a) Simple apparatus for reflux (b) Apparatus for reflux with drying device (c) Apparatus for reflux with gas-absorbing device

Figure 1-3 Reflux Apparatus

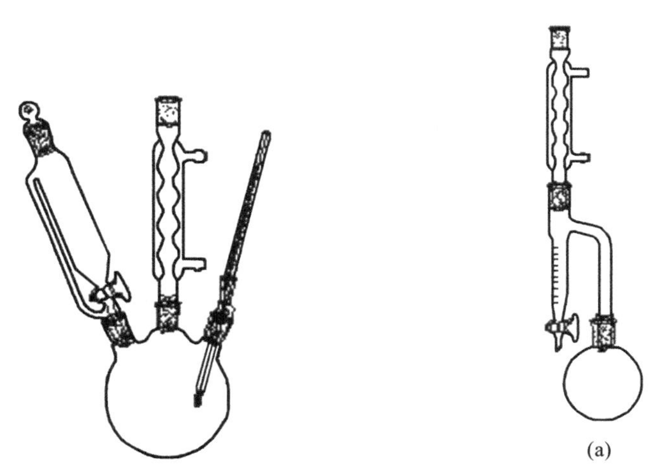

Figure 1-4 Addition Reflux Apparatus

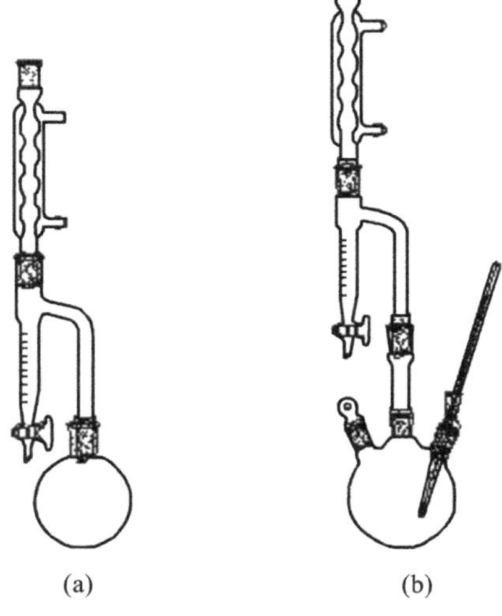

(a) (b)

Figure 1-5 Water Diversion Reflux Apparatus

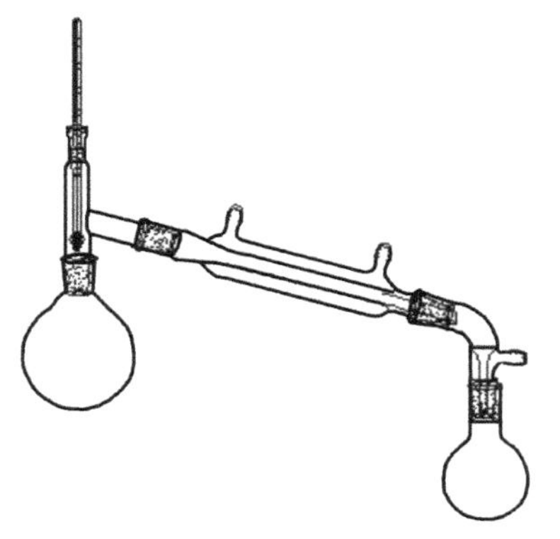

(a) Water condensing apparatus for distillation

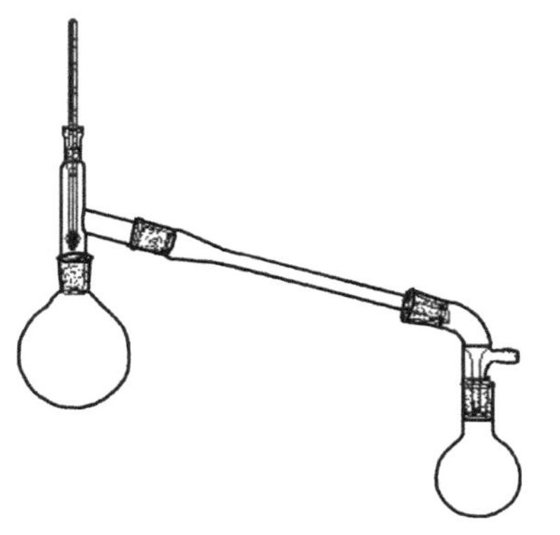

(b) Air condensing apparatus for distillation

Figure 1-6 Simple Distillation Apparatus

3. Fractional distillation apparatus

Fractional distillation apparatus is shown in Figure 1-8.

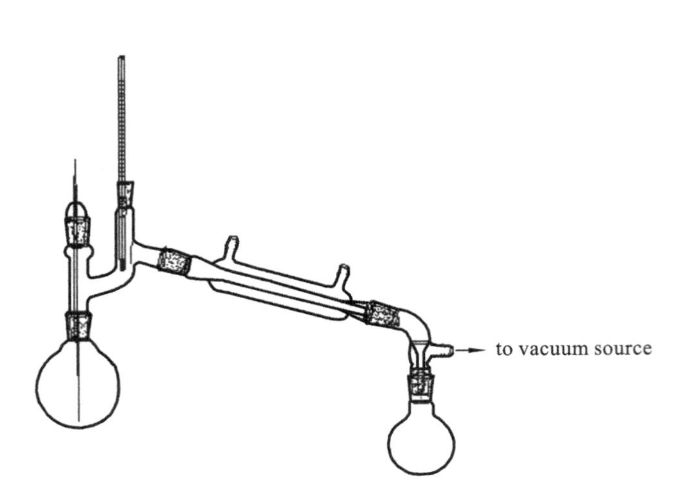

Figure 1-7　Vacuum Distillation Apparatus

Figure 1-8　Fractional Distillation Apparatus

Ⅲ Notes of Glassware Usage

(1) All should be used carefully, avoiding impact or breakage.

(2) Don't heat glassware directly except boiling flasks and tubes.

(3) Erlenmeyer flask and flat-bottom flask cannot withstand reduced pressure and should not be used in such systems.

(4) After cleaning up glassware containing a stopper, a small piece of paper must be put between the stopper and ground joint to avoid adhesion in next use.

(5) The glass of a mercury bulb is thin and easy-to-break thus should be used with care. Never use it as stirring rod! After use, cool it down, and rinse it afterwards to keep away from cracking. The measurement of thermometer doesn't go beyond its graduated range.

Ⅳ Notes of Apparatus Assemble

(1) All glassware and accessories must be clean and fitted properly.

(2) When assembling the apparatus, follow the principle of "bottom-to-top, left-to-right", one by one. Clamp at the joint and never clamp too tightly.

(3) When disassembling, observe the rule of "right-to-left, up-to-down", one by one. The disassembly should be done after the apparatus cool down and no condensate appears.

(4) A reaction apparatus under the ordinary pressure must have an opening to the atmosphere to avoid development of a dangerously high pressure within the system when heat is applied.

(5) All experimental apparatus must be tight, right, tidy and safe. All ground-glass joints should be made connected snugly.

Ⅴ Cleaning Glassware

Always wash your glassware at the end of the experiment with water and either detergent or mild scouring powder. Using an appropriate brush to remove most organic chemicals adhered to the glass wall. The inside and outside of all pieces of apparatus should be scrubbed, and rinsed thoroughly with water afterwards. The final rinse can also be done with distilled or deionized water as required. Sometimes an ultrasonic oscillator might be useful for cleaning.

Never use chemical reagents or organic solvents thoughtlessly to rinse glassware. This may produce wastes, and create a hazardous situation, resulting in additional pollution to our environment.

Ⅵ Drying Glassware

The common methods of drying glassware are as follows:

(1) Air dry: In order to let water stream down, the glassware can be left upside down on a drying rack to dry.

(2) Oven dry: The glassware can be dried quickly by placing it in an oven. For complete drying, glass should be left in an oven at 110～120 ℃ for several hours. Besides this, an airflow drier or hair drier also can be used.

(3) Organic solvent dry: When wet glassware must be dried quickly for immediate use, it may be rinsed with small amounts of organic solvent such as 95% ethanol or acetone, which must be drained into an assigned bottle after use. Use hair drier to evaporate solvent afterwards.

1.5 Common Organic Solvents—Properties and Treatments

Organic solvents must be handled safely. Always remember that organic solvents are all at least mildly toxic and that many are flammable. The most common organic solvents are listed below with their important physical constants and properties. You should become thoroughly familiar with safety caution on their treatments.

Ⅰ Ethyl Ether ($C_2H_5OC_2H_5$)

Additional names: Diethyl ether; ether; ethoxyethane; ethyl oxide; anesthetic ether.

Molecular weight: 74.12.

Physical constants: Density 0.7134; b. p. 34.6 ℃; flash point −49 ℃; n_D^{20} 1.351～1.353; water solubility (20 ℃) 6.9 g · $(100\ mL)^{-1}$.

Properties: Extremely flammable liquid with characteristic, sweetish, pungent odor; tend to form explosive peroxide under the influence of air and light when evaporation to dryness is attempted. Peroxides may be removed from ether by shaking with 5% ferrous sulfate aqueous solution. Addition of naphthols, polyphenols, aromatic amines, and aminophenols has been proposed for the stabilization of ethyl ether. Ether is slightly soluble in water and water is slightly soluble in ether. Be miscible with lower aliphatic alcohols, benzene, chloroform, other organic solvents and many oils.

Safety caution: Keep away from fire; do not empty into drains. Vapors may cause drowsiness and dizziness, repeated exposure may cause skin dryness or cracking.

Ⅱ Ethyl Alcohol (C_2H_5OH)

Additional names: Ethanol; absolute alcohol; anhydrous alcohol; dehydrated alcohol; alcohol; distilled spirits.

Molecular weight: 46.07.

Physical constants: Density 0.789; b. p. 78.5 ℃; flash point 12 ℃; n_D^{20} 1.3614.

Properties: Clear, colorless, highly flammable liquid with pleasant odor; be miscible with water and with many organic liquids. Absorb water rapidly from air.

Safety caution: Keep container tightly closed; keep away from flame!

Ⅲ Acetone (CH_3COCH_3)

Additional names: 2-Propanone; dimethyl ketone; beta-ketopropane; methyl ketone; propanone; pyroacetic acid; pyroacetic ether.

Molecular weight: 58.08.

Physical constants: Density 0.788; b. p. 56.5 ℃; Flash point −20 ℃; n_D^{20} 1.3591.

Properties: Volatile, highly flammable liquid; characteristic odor; pungent, sweetish taste. Be miscible with water, alcohol, dimethylformamide, chloroform, ether.

Safety caution: Keep container in a well-ventilated place. Keep away from sources of ignition. Keep away from plastic eyeglass frames, jewelry, pens and pencils, rayon stockings and other rayon garments. In case of contact with eyes, rinse immediately with plenty of water and seek medical advice.

Ⅳ Benzene (C_6H_6)

Additional names: Benzol; benzolene.

Molecular weight: 78.11.

Physical constants: Density 0.8786; b. p. 80.1 ℃; flash point −11 ℃; n_D^{20} 1.5011; water solubility (20 ℃) 0.18 g·(100 mL)$^{-1}$.

Properties: Clear, colorless, volatile, highly flammable liquid; characteristic odor; slightly soluble in water. Be miscible with alcohol, chloroform, ether, carbon disulfide, carbon tetrachloride, glacial acetic acid, acetone, oils.

Safety caution: Keep in well-closed containers in a cool place and away from fire. Direct contact may cause irritation of eyes, respiratory system and skin; aspiration into the lung may lead to chemical pneumonitis. Prolonged exposure may cause cancer and heritable genetic damage.

Ⅴ Ethyl Acetate ($CH_3COOC_2H_5$)

Additional name: Acetic acid ethyl ester.

Molecular weight: 88.11.

Physical constants: Density 0.902 (0.898 g/cm^3, 25 ℃); b.p. 77 ℃; n_D^{20} 1.3720; water solubility (20 ℃) 0.18 g · (100 mL)$^{-1}$.

Properties: Clear, volatile, highly flammable liquid; characteristic fruity odor. Slowly decomposed by moisture, then acquires an acid reaction. Be slightly soluble in water. Be miscible with alcohol, acetone, chloroform, ether.

Safety caution: Keep tightly closed in a cool place and away from flame; irritating to eyes. Repeated exposure may cause skin dryness or cracking. Vapors may cause drowsiness and dizziness. Take precautionary measures against static discharges.

Ⅵ Trichloromethane ($CHCl_3$)

Additional names: Chloroform; formyl trichloride; methane trichloride.

Molecular weight: 119.38.

Physical constants: Density 1.484; b.p. 61～62 ℃; n_D^{20} 1.4476.

Properties: Highly refractive, nonflammable, heavy, very volatile, sweet-tasting liquid; characteristic odor. Slightly soluble in water; Miscible with alcohol, benzene, ether, petroleum ether, carbon tetrachloride, carbon disulfide, oils.

Safety caution: Potential symptoms of overexposure are dizziness, mental dullness, anesthesia; limited evidence of a carcinogenic effect. Direct contact may cause irritation to eyes and skin. Prolonged exposure may cause danger of serious damage to health. Wear suitable protective clothing or suitable gloves when handle with it.

Ⅶ *n*-Hexane (C_6H_{14})

Molecular weight: 86.18.

Physical constants: Density 0.659; b.p. 69 ℃; flash point −22℃; n_D^{20} 1.3748.

Properties: Colorless, very volatile, highly flammable liquid; faint, peculiar odor. Be insoluble in water. Be miscible with alcohol, chloroform, ether.

Safety caution: Keep away from sources of ignition. Toxic to aquatic organisms. Potential symptoms of overexposure are light-headedness; nausea, headache; numbness of extremities, muscle weakness; irritation of eyes and skin. Wear suitable protective clothing or suitable gloves when handle with it. Do not empty into drains. Avoid release to the environment. If swallowed, do not induce vomitting, seek medical advice immediately and show this container or label.

Ⅷ Petroleum Ether and Ligroin

Petroleum ether is a mixture of hydrocarbons with isomers of formulas C_5H_{12} and C_6H_{14} predominating. Petroleum ether is often referred to the one has boiling point ranging from about 30 ℃ to about 60 ℃. Note that petroleum ether is not an ether at all because there are no oxygen-bearing compounds in the mixture.

Ligroin, or high-boiling petroleum ether, is like petroleum ether in composition except that ligroin generally includes higher-boiling alkane isomers which have boiling points ranging from about 60 ℃ to about 90 ℃ or from about 90℃ to about 120℃.

Ⅸ Tetrahydrofuran (C_4H_8O)

Additional names: THF; diethylene oxide; tetramethylene.

Molecular weight: 72.11.

Physical constants: Density 0.8892; b. p. 66 ℃; flash point −14℃; n_D^{20} 1.4070.

Properties: Clear, colorless, highly flammable, viscous liquid; ether-like odor. Be miscible with water, alcohols, ketones, esters, ethers, and hydrocarbons.

Safety caution: May form explosive peroxides. Distil only in presence of a reducing agent, such as ferrous sulfate. Overexposure may cause irritation of eyes and upper respiratory system; nausea, dizziness and headache. Do not empty into drains. Take precautionary measures against static discharges.

Ⅹ N,N-Dimethylformamide (C_3H_7NO)

Additional names: DMF; formdimethylamide; dimethyl formamide.

Molecular weight: 73.09.

Physical constants: Density 0.9445; b. p. 153 ℃; flash point 58 ℃; n_D^{25} 1.4290.

Properties: Colorless or very slightly yellow liquid; faint amine odor; miscible with water and most common organic solvents.

Safety caution: Harmful by inhalation; harmful in contact with skin. May cause harm to the unborn child. Avoid exposure, obtain special instructions before use. In case of accident or if you feel sick, seek medical advice immediately (show the label whenever possible.)

Ⅺ Pyridine (C_5H_5N)

Molecular weight: 79.10.

Physical constants: Density 0.9827; b. p. 115.2 ℃; flash point 17 ℃; n_D^{20} 1.5085～1.5105.

Properties: Highly flammable, colorless liquid; characteristic disagreeable odor; sharp taste. Be miscible with water, alcohol, ether, petroleum ether, oils and many other organic liquids. It's weak base, forms salts with strong acids.

Safety caution: Potential symptoms of overexposure are headache, nervousness dizziness and insomnia; nausea, anorexia; frequent urination; eye irritation; dermatitis; liver and kidney damage. After contact with skin, wash immediately with plenty of soap-suds.

Ⅻ Acetonitrile (CH_3CN)

Additional names: Methyl cyanide; cyanomethane.

Molecular weight: 41.05.

Physical constants: Density 0.7875; b. p. 81.6 ℃; flash point 2 ℃; n_D^{15} 1.3460.

Properties: Highly flammable, clear, colorless liquid; ether-like odor. Be soluble in water; miscible with methanol, methyl acetate, ethyl acetate, acetone, ether, acetamide solutions, chloroform, carbon tetrachloride, ethylene chloride and many unsaturated hydrocarbons.

Safety caution: Poisonous! Direct contact may cause skin and eye irritation. Wear suitable protective clothing and gloves when handle with it.

1.6 The Writing of Preview and Final Reports

Ⅰ Advance Preparation

Every student should prepare a lab notebook and must carefully and fully preview each experiment before start. The specific requirements of the preview are as follows:

(1) Preview what are requested by the instructions for each experiment in this book.

(2) Understand the reaction principle (main reaction, and the side reaction).

(3) Write physical constants of main reagents and products, and specifications and quantities (g, mL, mol) of the main reagent; list all glassware needed.

(4) List general procedures of the experiment and any chemical hazards it might present.

(5) List a table for recording the data.

Ⅱ Lab Records

During experiment, the student should operate formally, observe carefully and think actively, and also should record timely the experimental phenomena (such as the heating condition, the change of the color or pH, precipitation and gas emitting, etc.) and data (such as room temperature, reaction temperature, etc.). Writing notes and recording results by later memories are never allowed. The lab record should be real, concise and clear. After the experiment is finished, the original records and the products obtained should be presented and checked by teacher before leave.

Ⅲ Experiment Reports

Experiment report is an important part of experimental practice for a student. A complete final report should include the objectives (what to do), the principle (why to do), the procedures with the reagents and apparatus you have used (how to do), and the results you have obtained. The final summary and conclusion on your experimental results are especially important and the discussions of improvement advices or solutions to some experimental failures are also required. A report sample is given as follows:

Sample:Experiment Title ______________________

1) Objectives of the experiment: Main techniques introduced for the experiment

2) Experimental principle: Chemical reaction; balanced chemical equation or reaction mechanism

3) Physical constants of main reagents and the product

4) The dosage and specification of main reagents

5) Experimental apparatus

6) Experimental procedures and observations: Color change, solubility, precipitation etc.

7) Calculation of yield percentage

8) Conclusion and discussion

第二部分　基本实验技能

Part 2　Basic Experimental Skills Training

实验一　加热与冷却

一、加热

在室温下，某些反应难以进行或反应速率很慢。为了提高反应速率，需要在加热的条件下进行反应。此外，有机化合物的蒸馏、升华等操作也都需要加热。有机化学实验室常用的热源有煤气灯、电热套、电热板等。加热的方式有直接加热和间接加热，但在有机化学实验室一般不采用直接加热，因为玻璃器皿会因受热不均匀而破裂。同时，局部过热还可能导致某些有机化合物的分解，甚至燃烧或爆炸。为了避免直接加热可能带来的安全问题，保证加热均匀，有机化学实验室常根据具体情况采用下列热浴进行间接加热。

1. 水浴

水浴是一种加热均匀、温度易控制的加热方式，当加热温度不超过 100 ℃时，最好用水浴（见图 2-1）加热。加热温度在 90 ℃以下时，可将盛物料的容器部分浸入水中（注意勿使容器接触水浴装置底部），调节热源的大小，把水温控制在所需范围以内。如需加热至 100 ℃，可用沸水浴；也可将容器悬空置于水浴上方，利用水蒸气来加热。如欲停止加热，只需将浴底的热源移开，水即停止沸腾，容器的温度就会很快地下降。

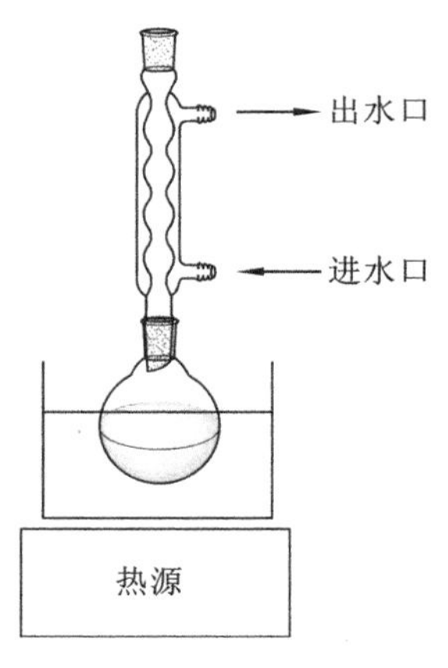

图 2-1　水浴加热回流装置

需要注意的是，由于较长时间加热，水浴中的水会不断蒸发，因此在操作时应适当补充热水，以使水浴中水面始终略高于容器内的液面。

2. 油浴

加热温度在 100～250 ℃时，可用油浴。油浴的优点在于温度易控制在一定范围内，容器内的反应物受热均匀。容器内反应物的温度一般比油浴温度低 20℃左右。与水浴加热一样，油浴中导热油的液面始终略高于容器内的液面。

用油浴加热时，要特别当心，防止着火。当油的冒烟情况严重时，应立即停止加热。万一着火，应首先关闭电源或煤气灯，再移去周围易燃物，然后用石棉布盖住油浴口，火即可熄灭。油浴中应悬挂温度计，以便随时调节电压或灯焰，控制温度。油浴中的油量不能过多，否则受热后有溢出而引起火灾的危险；导热油中也不能溅入水，否则加热时会产生泡沫或引起爆溅。

加热完毕后，将容器提离油浴液面，仍用铁夹夹住，悬空置于油浴上方。待附着在容器外壁的导热油滴完后，用纸或干布擦净。

油浴能达到的最高温度取决于所使用油的种类，常用的导热油有以下几种。

(1) 甘油，可加热至 140～150 ℃，温度过高会发生分解，并释放出难闻的气味。甘油吸水性强，长时间放置的甘油在使用前应首先加热除去所吸收的水分。

(2) 植物油，如菜油、蓖麻油、花生油等，可加热至 220 ℃。使用时常加入 1%的对苯二酚等抗氧化剂，以增加其热稳定性。该类导热油在温度过高时会分解，达到闪点时会发生燃烧，因此使用时应小心。

(3) 石蜡，可加热至 200 ℃左右，优点是冷却至室温后会凝成固体，便于储藏，因此在加热完毕后，应及时从石蜡浴中取出容器。

(4) 液状石蜡，可加热至 220 ℃左右，较高温度下不易分解，但较易燃烧。

(5) 硅油，加热至 250 ℃时仍较稳定，透明度好，但价格较贵。

3. 空气浴

空气浴是利用热空气进行间接加热，对于沸点在 80 ℃以上的液体均可采用。将容器置于石棉网上，两者间隔约 1 cm，利用煤气灯隔着石棉网对容器进行加热，这就是最简单的空气浴。但是这种方法加热较猛烈，受热仍不均匀，故不能用于回流沸点较低、易燃的液体或者减压蒸馏。

除煤气灯外，有机化学实验室经常采用电热套作为空气浴的热源(见图 2-2)。由于电热套中的电热丝是用玻璃纤维包裹着的，所以更为安全，一般能从室温加热至 200 ℃左右。电热套主要用于回流加热，蒸馏或减压蒸馏时不用为宜，因为在蒸馏过程中随着容器内物质逐渐减少，会使容器壁过热。安装电热套时，应使容器外壁与电热套内壁保持约 2 cm 的距离，以便利用热空气传热，并防止局部过热。

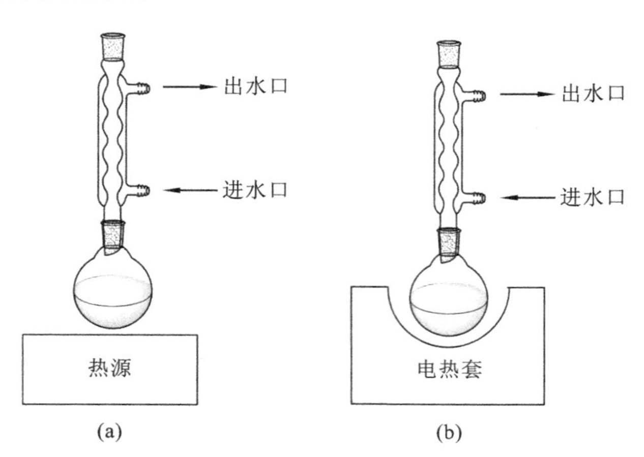

图 2-2　空气浴加热回流装置

4. 沙浴

沙浴使用方便，可加热至 350 ℃。一般将清洁而干燥的细沙盛于铁盘上，将容器半埋在沙中加热。沙浴的缺点是对热的传导能力较差而散热却较快，导致沙浴温度分布不均，且不易控制。因此容器底部的沙层略薄，使容器易受热；而容器周围的沙层略厚，使热量不易散失。沙浴中应插入温度计，以便控制温度，且温度计的水银球应紧靠容器。使用沙浴时，桌面应铺石棉布，以防辐射热烤焦桌面。

二、冷却

放热反应进行时，常伴随大量的热产生，使反应温度迅速升高。如果控制不当，往往会引起反应物的蒸发，逸出反应器，也可能引发副反应，甚至引起爆炸。为了将温度控制在一定范围内，需要适当进行冷却。因此，在有机化学实验中，经常需要采用一定的冷却剂进行冷却操作，在一定的低温条件下进行反应、分离、提纯等。

冷却剂的选择是根据冷却需要达到的温度和带走的热量来进行的。

1. 水

水是所有冷却剂中成本最低的，且热容量高，故为常用的冷却剂。但随着季节的不同，其冷却效率变化较大。此外，在回流、蒸馏等操作中也常用水做冷却剂，以冷却高温的气体或液体。

2. 冰-水混合物

某些反应需在低于室温的条件下进行，则可用水和碎冰的混合物作冷却剂，冰-水浴可冷却至 0～5 ℃。由于能和容器更好地接触，其冷却效果要比单用冰块好。如果水的存在并不妨碍反应的进行，则可以把碎冰直接投入反应物中，这样能更有效地保持低温。

3. 冰-盐混合物

如果需要将反应混合物保持在 0 ℃以下，常用碎冰和无机盐的混合物作冷却剂。制作冰-盐浴时，应把盐研细，然后和碎冰按一定比例均匀混合。混合比例及冰盐浴能达到的最低温度参见表 2-1。在实验室中，最常用的冷却剂是碎冰和食盐的混合物，一般能将反应物冷却至－21～－5 ℃。

表 2-1　常用冰-盐浴冷却剂

盐　　类	100 g 碎冰中加入盐的质量/g	混合物能达到的最低温度/℃
NH_4Cl	25	－15
$NaNO_3$	50	－18
NaCl	33	－21
$CaCl_2 \cdot 6H_2O$	100	－29
$CaCl_2 \cdot 6H_2O$	143	－55

4. 干冰(固体二氧化碳)

干冰可冷却至－60 ℃以下。如果将干冰加到乙醇、丙酮等溶剂中，可冷却至－78 ℃，但加入时会猛烈起泡。使用该类冷却剂时，应将冷却剂置于杜瓦瓶(广口保温瓶)中或其他绝热效果好的容器中，以保持其冷却效果。

5. 液氮

液氮可冷却至－196 ℃。为了保持冷却效果，应将这种冷却剂置于杜瓦瓶(广口保温瓶)中或其他绝热效果好的容器中。

应当注意，当冷却温度低于－38 ℃时，不能使用水银温度计，因为水银的凝固点为－38.9 ℃，测定温度时应采用内部添加少许颜料的有机液体(如乙醇、甲苯、正戊烷等)低温温度计。

Experiment 1　Heating and Cooling

Ⅰ Heating

At room temperature, some organic reactions are difficult to carry out or the reaction rate is very slow. In order to improve the reaction rate, the condition of heating is generally required. The processes of distillation and sublimation all require heating. Several heat sources, such as Bunsen burner, heating mantle and electric hot plate, are commonly used in the organic chemistry lab. Usually, heating is not carried out straightforwardly. To avoid possible safety problems from straightforward heating, the following indirect heating methods are generally used according to the specific circumstances.

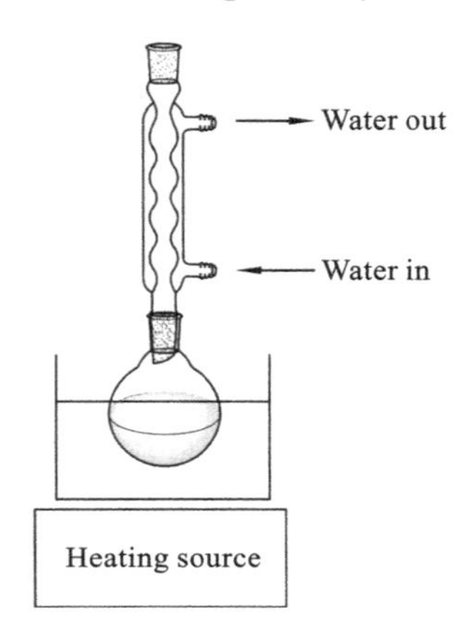

Figure 2-1　Apparatus for Reflux on a Water Bath

1. Water bath

In the case of solutions of flammable liquids having a boiling point below 100 ℃, the electrically-heated water bath provided with a constant-level device must be used(see Figure 2-1). If the heating temperature is below 90 ℃, the vessel can be immersed into the water bath directly (without permission to touch the bottom of the bath). The temperature of 100 ℃ can be achieved by using boiling water bath or steam bath. In the operation of steam bath, the vessel is suspended above the water, which is heated by the vapor.

Due to the evaporation of the water, a certain amount of hot water should be added during the operation, so that the surface of water bath is kept higher than that of solution inside.

2. Oil bath

For temperatures in the range of 100 ~ 250 ℃, oil bath is generally used. The temperature is easily controlled in a certain range, and the reactant in the vessel is heated evenly. The reaction temperature is generally lower than that of oil bath about 20 ℃. It is same to water bath that the surface of oil bath should be kept higher than that of solution inside.

The operators must take particular care to prevent fire when using oil bath. The heating should be immediately stopped when the oil is fuming severely. Once catching fire, the operators ought to turn off the power or Bunsen burner firstly, and then remove the surrounding combustibles, finally cover the oil bath with asbestos. A thermometer should always be placed in the bath to avoid excessive heating. Introducing water into the bath is also not permitted, which may splatter from the hot oil. Flasks, when removed from an oil bath, should be allowed to drain for several minutes above the bath and then wiped with a rag.

The highest temperature that oil bath can reach is dependent upon what kind of oil is used. Several heat conducting oils, commonly used in the organic chemistry lab, are presented as follows:

(1) Glycerol is satisfactory up to 140～150 ℃. Above these temperatures, decomposition is usually excessive and the odor of the vapors is unpleasant. Due to its high hydrophilicity, glycerol should be heated to remove the absorbed moisture before use when laying aside for a long time.

(2) For temperatures up to about 220 ℃, plant oil, such as rapeseed oil, castor oil and peanut oil, is recommended. One percent of hydroquinone or other antioxidant is usually added to improve the thermal stability of the oil. It should be noted that plant oil easily decomposes at high temperature and combustion occurs when reaching its flash point.

(3) Paraffin can be heated to about 200 ℃. It is convenient for storage because the hot oil will turn into a solid when cooling down to room temperature. Therefore, paraffin should be taken out from the bath immediately after use.

(4) Liquid paraffin can be heated up to about 220 ℃. It is of high thermal stability, but easier to burn at high temperature.

(5) Silicone fluid is probably the best liquid for oil bath but somewhat expensive for general use. It may be heated high up to 250 ℃ without appreciable decomposition and discoloration.

3. Air bath

Air bath is a very cheap and convenient method of effecting heating of small distillation flasks, which is satisfactory for the liquids having a boiling point above 80 ℃. When using Bunsen burner as heating source, it is essential to use a heat resistant bench mat, which is away from the flask about 1 cm. This is the simplest air bath. However, this method easily results in fluctuations at the level of heating due to air draughts. It is not suitable for the reflux of liquids with low boiling point or distillation under reduced pressure.

In the organic chemistry lab, heating mantle is the most commonly used as heating source for air bath (see Figure 2-2). They consist of a heating element enclosed within a knitted glass-fiber, which makes it much safer. Heating mantle is satisfactory from room temperature to about 200 ℃. This method is mainly used for reflux, preferably not for simple distillation and vacuum distillation, which may be attributed to the partly overheating of the flask with the decreasing of the liquid. Therefore, the flask should be kept away from the heating mantle about 2 cm to avoid overheating.

4. Sand bath

When the required heating temperature is much higher than those listed above, one can often use a sand bath. Its highest operating temperature is up to 350 ℃. The clean and dry sands are placed in iron plate, in which the vessel is half buried. The disadvantage of the sand bath is the poor heat conductivity, but fast heat dissipation, resulting in uneven distribution of the temperature. Hence the sands in the bottom of the plate should be slightly thinner, but

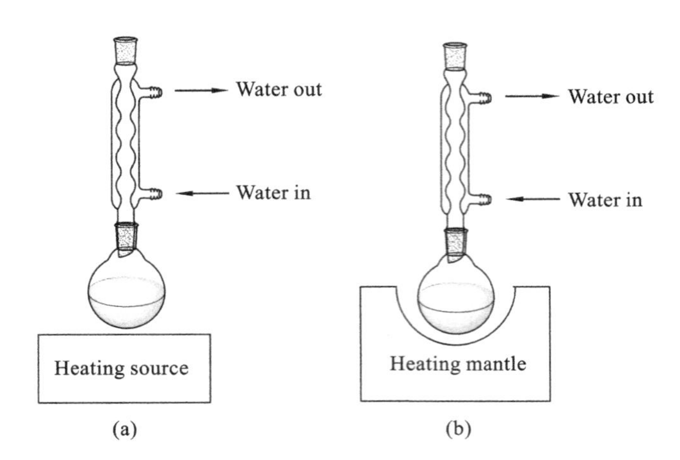

Figure 2-2　Apparatus for Reflux on an Air Bath

slightly thicker around the vessel. A thermometer should always be inserted in the bath to detect the temperature, of which the mercury bulb should be close to the vessel. The asbestos should also be used to protect the bench.

Ⅱ Cooling

Exothermic reactions are usually accompanied by a large heat generation, which increases the reaction temperature rapidly. This tends to cause evaporation of the reactants and side reactions, even an explosion. It is often necessary to obtain low temperature for these reactions by immersing the reaction vessel in a cooling bath. This method is also used in the processes of separation and purification.

The coolants should be selected according to the low temperature to be achieved and quantity of heat to be taken away. Several main types of coolants are described below.

1. Water

Water is the most commonly used coolant because of its lowest cost and high thermal capacity. It should be noted that the cooling efficiency greatly changes with the different seasons. Moreover, water is often used in reflux and distillation to cool the hot gas or liquid.

2. Ice-water mixture

Finely crushed ice is used for maintaining the temperature at 0～5℃. It is usually best to use a slush of crushed ice with water to provide sufficient contact with the vessel to be cooled. If the water does not interfere with the reaction, crushed ice can be directly put into the reaction mixture to maintain low temperature effectively.

3. Ice-salt mixture

For temperatures below 0℃, the commonest freezing mixture is a mixture of crushed ice and inorganic salt in certain ratio. The lowest temperatures that ice-salt bath can reach with different ice-salt ratios are shown in Table 2-1. A mixture of one part of sodium chloride and three parts of ice is the most commonly used in the lab, which can produce a temperature of −21 to −5 ℃.

Table 2-1 Common Ice-salt Mixtures

Salt	Mass of salt per 100 g of crushed ice/g	Lowest temperature approx. /℃
NH_4Cl	25	−15
$NaNO_3$	50	−18
NaCl	33	−21
$CaCl_2 \cdot 6H_2O$	100	−29
$CaCl_2 \cdot 6H_2O$	143	−55

4. Dry ice

Dry ice is employed to obtain the temperature below −60 ℃. If small pieces of dry ice are carefully added to either ethanol or acetone until the lumps of dry ice no longer evaporates vigorously, the temperatures attained are low to −78 ℃. In order to keep the freezing mixture for hours or overnight, it should be prepared in a Dewar flask.

5. Liquid nitrogen

The attainment of temperatures lower than −100 ℃ requires the use of baths employing liquid nitrogen, which can be as cold as −196 ℃. It should also be stored in a Dewar flask to keep the cooling efficiency.

It should be noted that the mercury thermometer is forbidden to use when the temperature is below −38 ℃, because the freezing point of mercury is −38.9 ℃. It is preferably to choose the low temperature thermometer based on ethanol, toluene or *n*-pentane with a little pigment.

实验二 有机物的干燥与无水乙醇的制备

一、实验目的

(1) 了解有机化合物干燥的基本原理，以及干燥剂的种类和使用方法。

(2) 掌握制备无水乙醇的原理和方法。

(3) 掌握干燥管的装法。

二、实验原理

有机化合物在进行波谱分析、定性或定量化学分析之前，以及固体有机物在测定熔点前，都必须完全干燥，否则会影响结果的准确性；液体有机物在蒸馏前也需要进行干燥，除去其中的水分，否则前馏分较多，测得的沸点也会不准确。此外，许多有机化学反应需要在“绝对”无水条件下进行，因此既要干燥原料、溶剂和容器，还要在反应进行时隔绝空气中的湿气。所以在有机化学实验中，试剂和产品的干燥是非常普通且十分重要的基本操作。干燥的方法大致有物理方法和化学方法两类。物理方法主要有吸附、分子筛脱水等；化学方法则是用干燥剂除水，根据除水原理的不同又可分为与水结合生成水合物(如氯化钙、硫酸镁和硫酸钠等)和与水起化学反应(如五氧化二磷、氧化钙等)两种。

1. 液体有机物的干燥

1) 干燥剂的选择

液体有机物的干燥通常是将干燥剂直接加入有机物中,因此选择干燥剂时要考虑以下因素:不与被干燥的有机物发生化学反应;不能溶于该有机物中;吸水量大、干燥速度快、价格便宜。下面介绍几种最常用的干燥剂。

(1) 无水氯化钙。由于其吸水能力大(在 30 ℃以下形成 $CaCl_2 \cdot 6H_2O$,吸水容量为 0.97),价格便宜,因此在实验室中被广泛使用。但它吸水不快,因而干燥的时间较长。工业上生产的氯化钙往往含有少量的氢氧化钙或氧化钙,因此氯化钙不能用于酸性有机物的干燥;此外,由于氯化钙还能和醇、酚、酰胺、胺以及某些醛和酯形成配合物,因此氯化钙也不能用于上述有机物的干燥。

(2) 无水硫酸镁。它是一种适用范围较广的中性干燥剂,价格不太贵,干燥快,可以干燥不能用无水氯化钙干燥的有机化合物,如醛、酯等。其吸水容量为 1.05(按 $MgSO_4 \cdot 7H_2O$ 计)。

(3) 无水硫酸钠。它也是一种适用范围较广的中性干燥剂,其吸水容量为 1.25(32.4 ℃以下形成 $Na_2SO_4 \cdot 10H_2O$)。但其水合物具有较大的蒸气压(25 ℃时为 255.98 Pa),故干燥效能差。因此,无水硫酸钠常用于含水量较多的液体有机物的初步干燥,残留的水分再用干燥效能强的干燥剂进一步干燥。

(4) 无水碳酸钾。碳酸钾为弱碱性干燥剂,可用于醇、酮、酯等的干燥,但不能用于酸、酚和其他酸性物质的干燥,其吸水容量仅为 0.2(按 $K_2CO_3 \cdot 2H_2O$ 计)。

(5) 氢氧化钠和氢氧化钾。这是一类能有效干燥胺类化合物的干燥剂。由于氢氧化钠(或氢氧化钾)能和很多有机化合物发生反应(如酸、酚、酯和酰胺等),也能溶于某些液体有机物中,所以该类干燥剂的适用范围有限。

(6) 氧化钙。它这适用于低级醇的干燥。由于氧化钙和氢氧化钙均不溶于溶剂,且对热稳定,因此干燥后可直接蒸馏。氧化钙的强碱性,使其不能用于酸性物质和酯的干燥,其中后者会发生水解反应。

(7) 金属钠。它适用于干燥乙醚、脂肪烃和芳烃等。这些有机化合物在用钠干燥前,需用无水氯化钙等干燥剂将其中的大量水分除掉。使用时,金属钠要用刀切成薄片,最好是用金属钠压丝机将钠压成细丝后投入液体有机物中,以增大钠和液体的接触面积。切记不能用钠干燥氯代的溶剂,否则会发生剧烈反应,甚至爆炸。

各类有机化合物常用的干燥剂列于表 2-2 中。

表 2-2 各类有机化合物常用的干燥剂

有机化合物	干 燥 剂
烃	氯化钙、金属钠
卤代烃	氯化钙、硫酸镁、硫酸钠
醇	碳酸钾、硫酸镁、硫酸钠、氧化钙
醚	氯化钙、金属钠
醛	硫酸镁、硫酸钠
酮	碳酸钾、氯化钙(用于干燥高级酮)

续表

有机化合物	干　燥　剂
酯	硫酸镁、硫酸钠、氯化钙、碳酸钾
硝基化合物	氯化钙、硫酸镁、硫酸钠
有机酸、酚	硫酸镁、硫酸钠
胺	氢氧化钠、氢氧化钾、碳酸钾

2）干燥剂的用量

干燥剂的用量是根据干燥剂的吸水容量以及水在有机物中的溶解度来估算的，一般用量要高于理论值。而分子结构也会影响干燥剂的用量，含亲水性基团的化合物用量稍多。干燥剂的用量要适当。用量少，则干燥不完全；用量过多，因干燥剂表面吸附而造成被干燥有机物的损失。

一般干燥剂用量为每 10 mL 液体加入 0.5～1 g 干燥剂。由于存在液体含水量差异、干燥剂质量和吸水能力差异、干燥剂颗粒大小以及干燥时的温度不同等因素，所以较难界定干燥剂的具体用量，上述用量仅供参考。

3）操作方法

干燥时选用的干燥剂颗粒不能太大，而粉状干燥剂在干燥过程中易形成泥浆状，导致分离困难，应慎重选择。实际操作时，一般先加入少量干燥剂，进行充分振荡、静置，如出现干燥剂附着器壁、相互黏结、摇动不易流动等，则说明干燥剂用量不足，应再补充干燥剂，直至新添加的干燥剂不结块、不粘壁，摇动时能自由流动，则说明所加干燥剂用量合适。如加入干燥剂后出现水相，必须用吸管将水吸出，然后再添加新的干燥剂。

干燥前液体有机物呈混浊状，干燥后变澄清，这可简单地作为水分基本除尽的标志。需要注意的是，温度越低，干燥剂的干燥效果越好，因此干燥应在室温下进行，而干燥后液体有机物需要蒸馏时，必须将干燥剂和液体分离。

2. 固体有机物的干燥

干燥固体有机化合物，主要是为了除去残留在固体中的少量低沸点溶剂，如水、乙醚、乙醇、丙酮、苯等。常用的干燥方法如下：

1）晾干

固体在空气中自然晾干是最简便、最经济的干燥方法。该方法适用于在空气中稳定、不易分解、不易吸潮的固体。操作时，将待干燥固体置于干燥的表面皿或滤纸上，摊开成薄薄的一层，再用另一张滤纸覆盖起来，放在空气中慢慢晾干。

2）烘干

对于热稳定性好、熔点较高的固体有机物，可将待干燥固体置于表面皿或蒸发皿中，放在水浴上、沙浴上或两层隔开的石棉网上层烘干，也可放在恒温烘箱中或红外灯下烘干。操作时，要注意防止过热，加热的温度切记不能超过该固体的熔点，以免固体变色或分解。

3）干燥器干燥

对于易吸潮，以及在高温下易分解或升华的固体有机物，可置于干燥器中干燥。干燥器分为普通干燥器、真空干燥器和真空恒温干燥器。

（1）普通干燥器（图 2-3(a)）。盖与缸身之间的平面经过磨砂处理，在磨砂处涂以真空脂，使之密闭。缸中有多孔瓷板，瓷板下面放置干燥剂，上面放置盛有待干燥固体的表面皿等。

(2) 真空干燥器(图 2-3(b))。它的干燥效率较普通干燥好。真空干燥器上有玻璃活塞，用以抽真空，活塞下端呈弯钩状，口向上，防止通向大气时，因空气流速过快将固体吹散。为防止干燥器在负压下爆裂，干燥器周围应以金属网或防爆布围住。解除干燥器内真空时，应缓慢打开活塞，以免空气流速过快。

(3) 真空恒温干燥器(图 2-3(c))。当在烘箱或真空干燥器内干燥效果欠佳时，则要使用真空恒温干燥器(也称干燥枪)，但仅适用于少量样品的干燥。如图 2-3(c)所示，将盛有样品的小船置于夹层内，连接上盛有五氧化二磷的曲颈瓶，然后减压，至最高真空度时，停止抽气，关闭活塞后加热溶剂(溶剂的沸点应低于待干燥样品的熔点)，利用蒸气加热夹层的外层，从而使样品在恒定的温度下被干燥。若所需干燥的样品量较大，可用真空恒温干燥箱。

上述干燥器使用的干燥剂应根据样品所含的溶剂来选择。例如，五氧化二磷吸收水，生石灰吸收水和酸，无水氯化钙吸收水和醇，氢氧化钠吸收水和酸，石蜡片吸收乙醚、氯仿、四氯化碳、苯等有机溶剂。

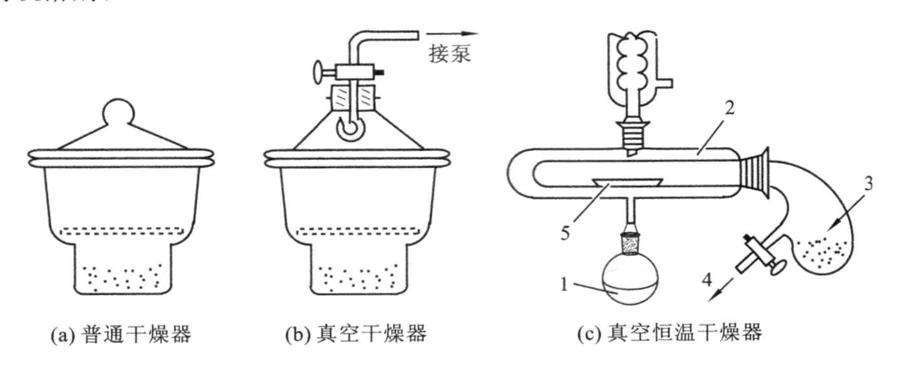

图 2-3　常用干燥器

1—溶剂；2—夹层；3—曲颈瓶中 P_2O_5；4—接真空水泵；5—盛样品的干燥舟

3. 无水乙醇的制备

在有机化学实验中经常需要用到高纯度的无水乙醇。一般而言，99.5%的无水乙醇基本可以满足实验的需求。该纯度无水乙醇可以从市场上直接购买，也可方便地由精馏乙醇经氧化钙干燥脱水制得。但不能由蒸馏法制得，因为 95.6%的乙醇和 4.4%的水可形成恒沸物。精馏乙醇即乙醇和水形成的恒沸物，具有恒定的沸点和体积比。加入氧化钙加热回流，可使乙醇中的水与氧化钙作用生成不挥发的氢氧化钙以除去水分。

$$CaO + H_2O \longrightarrow Ca(OH)_2 \downarrow$$

三、仪器与试剂

【仪器】圆底烧瓶，球形冷凝管，干燥管，蒸馏头，直形冷凝管，尾接管，三角烧瓶。

【试剂】精馏乙醇(125 mL)，氧化钙(35 g)，无水氯化钙。

【物理常数】

试剂	相对分子质量	熔点/℃	沸点/℃	密度/(g·cm^{-3})	溶解性	n_D^{20}
乙醇	46.07	−114.3	78.4	0.7893	与水混溶	1.3605
氧化钙	56.08	2572	2850	3.32～3.35	遇水反应	—

四、实验步骤

(1) 在 250 mL 圆底烧瓶中加入 125 mL 精馏乙醇和 35 g 氧化钙，依次安装球形冷凝管和氯化钙干燥管[1]。

(2) 加热回流 45 min。

(3) 停止加热，稍冷后改成蒸馏装置[2]，尾接管支管口安装氯化钙干燥管以隔绝湿气。蒸去前馏分(约 5 mL)后，换用另一已称量过的干燥接收瓶收集馏分。

(4) 称量所收集的无水乙醇[3]，并计算回收率。

五、注解和实验指导

【注解】

[1] 由于无水乙醇具有很强的吸水性，故操作过程中和存放时必须防止水分侵入。回流冷凝管顶端和尾接管上的氯化钙干燥管就是为了防止空气中的水分在回流及蒸馏的过程中进入体系内。干燥管的装法：在球端铺以少量棉花，然后在球部及直管部分加入适量颗粒状无水氯化钙，最后用棉花塞住顶端。

[2] 由于生成不挥发性的氢氧化钙，且加热不分解，故可留在瓶中一起蒸馏。

[3] 无水乙醇存放时应用空心塞塞住接收瓶口，以保证气密性。

【预习要求】

(1) 了解有机化合物干燥的基本原理，以及干燥剂的种类和使用方法。

(2) 了解回流、蒸馏装置的搭建方法，以及干燥管的装法。

【操作注意事项】

(1) 实验中所用仪器必须严格干燥。

(2) 所用乙醇的水分不能超过 0.5%，否则反应困难。

(3) 在回流、蒸馏过程中，以及存放无水乙醇时必须防止空气中的水分侵入。

(4) 回流时沸腾不宜过分剧烈，始终保持冷凝管下端有连续液滴即可。

六、思考题

(1) 用氧化钙制备无水乙醇的关键是什么？

(2) 本实验为何用氧化钙而不用无水氯化钙做无水乙醇的脱水剂？

(3) 本实验中，理论上需要氧化钙多少克？

Experiment 2　Drying Treatment of Organic Compounds and Preparation of Absolute Ethanol

Ⅰ Objectives

(1) To know the basic principles of the drying of organic compounds, and varieties and usage of the drying agents.

(2) To learn how to prepare absolute ethanol.

(3) To learn how to prepare a drying tube.

Ⅱ Principle

Before spectral analysis, qualitative or quantitative chemical analysis, and determination of the melting point, organic compounds must be completely dry, otherwise it will affect the accuracy of the results. Liquid organic compounds also need to be dried to remove the water prior to distillation, or else much more front distillates and inaccurate melting point will be obtained. In addition, many organic reactions require the absolutely anhydrous conditions. It is necessary to dry the reactants, solvents and vessels, and the reaction should be carried out in the absence of moisture as well. Therefore, drying of the organic compounds is a very useful and important technique in organic experiments. The common drying method can be primarily divided into physical and chemical ones. The physical drying method includes absorption, dehydration of molecular sieve, etc. The chemical drying method is to use drying agent to remove water, based on the principle of combination with water to form hydrates (e. g. calcium chloride, magnesium sulfate, sodium sulfate, etc.) and reaction with water directly(e. g. phosphorus pentoxide, calcium oxide, etc.).

1. Drying of liquid organic compounds

1) Selection of drying agents

Liquid organic compounds are usually dried by direct contact with a solid drying agent. The selection of a drying agent will be governed by the following considerations: (i) it must not react with the compound to be dried; (ii) it should be insoluble in the liquid; (iii) it should have a rapid action and an effective drying capacity; and (iv) it should be as economical as possible. The various common drying agents are discussed in detail below. Common drying agents for each kind of organic compounds are listed in Table 2-2.

(1) Anhydrous calcium chloride. This reagent is widely employed because of its high drying capacity and its cheapness. It has a high water-absorption capacity(0. 97, since it forms $CaCl_2 \cdot 6H_2O$ below 30 ℃) but is not rapid in its action. Ample time must therefore be given for drying. The industrial process for preparing the reagent usually permits a little hydrolysis to occur, and the product may contain some free calcium hydroxide or calcium oxide. It cannot therefore be employed for drying acids or acidic liquids. Calcium chloride can combine with alcohols, phenols, amides, amines and some aldehydes and esters, and thus cannot be used with these classes of compounds.

(2) Anhydrous magnesium sulfate. This reagent is an excellent neutral desiccant, rapid in its action and chemically inert, and thus can be employed for most compounds including those (aldehydes, esters, etc.) to which calcium chloride is not applicable. The water-absorption capacity of magnesium sulfate is 1. 05(the fully hydrated form is the heptahydrate).

(3) Anhydrous sodium sulfate. This is a neutral drying agent, is inexpensive, and has a high water-absorption capacity(1. 25, forming $Na_2SO_4 \cdot 10H_2O$ below 32. 4 ℃). It can be used on almost all occasions, but the drying action is slow and not thorough. This desiccant is valuable for the preliminary removal of large quantities of water. But sodium sulfate is useless above 32. 4 ℃, at which temperature the decahydrate begins to lose water of crystallisation.

(4) Anhydrous potassium carbonate. This drying agent possesses a moderate efficiency and water-absorption capacity(0.2, the dihydrate is formed). Because of its weak basicity, it can be applied to the drying of alcohols, ketones and esters, but cannot be employed for acids, phenols and other acidic substances.

(5) Sodium and potassium hydroxides. The use of these efficient reagents should usually be confined to the drying of amines. These bases react with many organic compounds(e. g. acids, phenols, esters and amides) in the presence of water, and are soluble in some liquid organic compounds so that their use as desiccants is very limited.

(6) Calcium oxide. This reagent is commonly used for the drying of alcohols of low molecular weight. Both calcium oxide and calcium hydroxide are insoluble in the solvents, stable to heat, and practically non-volatile, hence the reagent need not be removed before distillation. Owing to its high alkalinity, it cannot be used for acidic compounds and esters. The latter would undergo hydrolysis.

(7) Sodium metal. This reagent is applicable to the drying of ether, aliphatic and aromatic hydrocarbons. These compounds must be pre-dried by other desiccants(e. g. anhydrous calcium chloride) to remove most of the water before employing sodium. When using sodium as drying agent, sodium should be cut to sheets using a knife, preferably prepared to wires using a sodium press, in order to increase the contact area between sodium and liquid. Sodium should never be added to chlorinated solvents because a vigorous or explosive reaction could occur.

Table 2-2 Some Common Drying Agents for Organic Compounds

Organic Compound	Drying Agent
Hydrocarbon	$CaCl_2$, Na
Halohydrocarbon	$CaCl_2$, $MgSO_4$, Na_2SO_4
Alcohol	K_2CO_3, $MgSO_4$, Na_2SO_4, CaO
Ether	$CaCl_2$, Na
Aldehyde	$MgSO_4$, Na_2SO_4
Ketone	K_2CO_3, $CaCl_2$ (for senior ketones)
Ester	$MgSO_4$, Na_2SO_4, $CaCl_2$, K_2CO_3
Nitrocompound	$CaCl_2$, $MgSO_4$, Na_2SO_4
Organic acid, phenol	$MgSO_4$, Na_2SO_4
Amine	NaOH, KOH, K_2CO_3

2) Dosage of drying agents

The proper dosage of a drying agent, usually more than its theoretical value, can be figured out based on its water-absorption capacity and water solubility in the target organic solvent. In view of molecular structure, slightly more drying agents can be used for those compounds containing the hydrophilic group. The dosage of desiccant should be appropriate. The drying is incomplete if there is no sufficient desiccant, but loss of the organic compound results from too much desiccant owing to the surface adsorption.

The general dosage of drying agent is 0.5～1 g per 10 mL of liquid. Due to the difference in water content, quality and water-absorption capacity of the desiccants, particle size and temperature, it is difficult to determine the specific amount of drying agent. Hence the above dosage is for reference only. Just note that the amount varies from experiment to experiment.

3) Procedure

The desiccant particles selected should not be too large, and the powdered desiccant is easy to form a slurry-like solution, resulting in separation difficulties. In practical operation, one should firstly add a small amount of desiccant to the organic liquid, then swirl the solution and let it stand while observing the desiccant. If it attaches to the vessel, which is all clumped together and difficult to flow, a little more desiccant should be added until no formation of clumps for the newly added desiccant after shaking. If sufficient water is present to cause the separation of a small aqueous phase, it must be sucked out of the solution by a dropper and the solution treated with a fresh portion of the desiccant.

In general, the organic liquid is turbid before drying, and becomes clear after drying, which can be regarded as a sign of sufficiently drying. It should be noted that the lower the temperature, the better the drying efficiency. Hence the drying should be carried out at room temperature. At higher temperatures the vapor pressures above the desiccants become appreciable and the water may be returned to the liquid. The drying agents must be removed by filtration before distillation.

2. Drying of solid organic compounds

Drying of solid organic compounds is mainly to remove the small amount of solvent with low boiling point in the solid, such as water, ether, ethanol, acetone and benzene. The drying methods, commonly used in the organic chemistry lab, are presented as follows:

1) Air dry

The organic solid is naturally dried in the air, being the most convenient and economical method for drying. This method is applicable to the drying of solid, which is stable and non-hygroscopic in the air. The solid samples should be spread out on a dry watch glass or filter paper, then covered by another filter paper and allowed to air dry.

2) Oven dry

For the solid organic compound with high thermal stability and melting point, it can be dried more quickly by putting them in a watch glass or evaporation dish, then heated on a hot-water bath, in an oven or under an infrared lamp. It should be noted that the heating temperature must be lower than its melting point in order to avoid discoloration and decomposition.

3) Drying in a desiccator

For the solid organic compound with high hygroscopicity and a feature of decomposition or sublimation at high temperature, it can be dried in a desiccator. The desiccators are divided into ordinary (i. e. atmospheric pressure) desiccator, vacuum desiccator and thermostatic vacuum desiccator.

(1) Ordinary desiccator(see Figure 2-3(a)). The joint between the lid and the base is an interchangeable ground flange and this joint needs lubrication(e. g. with vacuum grease) to seal the desiccator before use. There is a porous porcelain plate in the desiccator, below which the drying agent is placed to absorb the solvent and above which a watch glass containing the solid to be dried is placed.

(2) Vacuum desiccator (see Figure 2-3 (b)). The drying efficiency of the vacuum desiccator is much better than that of the ordinary desiccator. It is fitted with a glass stopcock. In this case the air inlet to the desiccator terminates in a hooked extension which serves to ensure that the air flow is directed in an even upward spread to prevent dispersal of the sample when the vacuum is released. In use, all vacuum desiccator must be sited in an appropriately sized and totally enclosed wire-mesh desiccator cage. The desiccator implosion may occur at any time when it is under vacuum, and represents a serious hazard. When releasing the vacuum, the stopcock should be slowly opened to avoid the overquick air flow.

(3) Thermostatic vacuum desiccator(see Figure 2-3(c)). Frequently the water or other solvent is so firmly held that it cannot be completely removed in a vacuum desiccator at ordinary temperatures. Large quantities of sample(100 g upwards) must therefore be dried in a vacuum oven at higher temperatures, which is commercially available. For smaller amounts of sample, a convenient lab vacuum oven is the so-called "drying pistol". An interchangeable glass joint assembly is shown in Figure 2-3(c). The vapor from a boiling liquid in the flask rises through the jacket, surrounds the drying chamber(holding the sample in a sample boat) and returns to the flask from the condenser. The drying chamber is connected to the vessel containing the drying agent (e. g. phosphorus pentoxide), which is attached to a water aspirator.

The drying agent, used in the above desiccators, is dependent upon the solvent in the sample. For examples, phosphorus pentoxide for water, calcium oxide for water and acids, anhydrous calcium chloride for water and alcohols, sodium hydroxide for water and acids, paraffin wax slices for organic solvents (e. g. ether, chloroform, carbon tetrachloride and benzene), respectively.

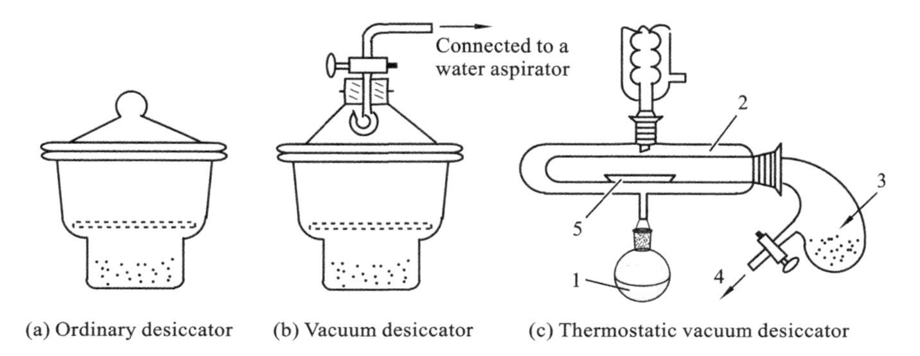

Figure 2-3 Some Common Desiccators

1—solvent; 2—jacket; 3—P_2O_5; 4—connected to a water aspirator; 5—drying boat containing sample

3. Preparation of absolute ethanol

Ethanol with a high degree of purity, which is the so-called "absolute ethanol", is frequently required in preparative organic chemistry. For some purposes absolute ethanol of about 99.5 percent purity is satisfactory, which is commercially available or conveniently prepared by the dehydration of rectified spirit with calcium oxide. But it cannot be prepared by distillation, owing to the formation of an azeotrope with 95.6 percent of ethanol and 4.4 percent of water. Rectified spirit is such an azeotrope with constant boiling point and volume ratio of the components. Calcium oxide is added into the rectified spirit, and the mixture is heated to reflux, which results in the reaction of calcium oxide with water. In this process, the involatile calcium hydroxide is produced.

$$CaO + H_2O \longrightarrow Ca(OH)_2 \downarrow$$

Ⅲ Apparatus and Reagents

【Apparatus】 round-bottom flask, Allihn condenser, drying tube, Liebig condenser, distilling head, and vacuum adapter.

【Reagents】 rectified spirit(125 mL), calcium oxide(35 g), anhydrous calcium chloride.

【Physical constants】

Reagent	M_w	m. p. /℃	b. p. /℃	$\rho/(g \cdot cm^{-3})$	Solubility	n_D^{20}
Ethanol	46.07	−114.3	78.4	0.7893	Miscible with water	1.3605
Calcium oxide	56.08	2572	2850	3.32～3.35	Reactive with water	—

Ⅳ Procedures

(1) Pour the rectified spirit(125 mL) and calcium oxide(35 g) into a 250 mL round-bottom flask. Fit the flask with an Allihn condenser, and then a calcium chloride drying tube[1].

(2) Reflux the mixture gently for 45 min.

(3) Stop heating, change to the distillation apparatus [2] after cooling down. Attach a calcium chloride drying tube to the vacuum adapter to isolate the ethanol from moisture. Collect the front distillate(about 5 mL), then use another weighed and dry receiver to collect the absolute ethanol.

(4) Weigh the collected absolute ethanol [3], and calculate the percentage yield.

Ⅴ Notes and Instructions

【Notes】

[1] Due to the high hydrophilicity of absolute ethanol, the moisture intrusion must be prevented during operation and storage. The apparatus is generally protected by a calcium chloride drying tube, which is fitted on top of the condenser, which is for reflux and attached to the vacuum adapter for distillation. The method for preparation of a drying tube is presented below: Place a small amount of cotton in the end of the bulb, then add some granular anhydrous calcium chloride in the bulb and straight tube, and finally plug the tube

with cotton.

[2] There is no need to remove the calcium hydroxide by filtration before distillation, owing to its nonvolatility and thermal stability.

[3] The vessel for the storage of absolute ethanol must be plugged with a glass stopper to ensure the air tightness.

【Requirements for preview】

(1) To be familiar with the basic principles of the drying of organic compounds, and varieties and usage of the drying agents.

(2) Get to learn the usage of apparatus for reflux, distillation and drying tube.

【Experimental precautions】

(1) All the glassware in this experiment must be completely dry.

(2) The water content should be less than 0.5 percent. Otherwise, the reaction is difficult to take place.

(3) The moisture intrusion must be prevented during reflux, distillation and storage.

(4) The boiling of the mixture should not be too severe. It is preferably to maintain the continuous drops at the bottom of the condenser.

Ⅵ Post-lab Questions

(1) What is the key to prepare absolute ethanol using calcium oxide?

(2) Why not anhydrous calcium chloride, but calcium oxide is used as drying agent for the dehydration of ethanol?

(3) How many grams of calcium oxide are theoretically required in this experiment?

Ⅶ Verbs

absolute ethanol 无水乙醇；
rectified spirit 精馏酒精；
calcium oxide 氧化钙；
anhydrous calcium chloride 无水氯化钙；
Liebig condenser 直形冷凝管；
drying tube 干燥管；
reflux 回流；
distillation 蒸馏

实验三　普通蒸馏及沸点测定

一、实验目的

(1) 了解普通蒸馏和沸点测定的原理及其应用。

(2) 掌握蒸馏烧瓶、冷凝管等的使用方法。

(3) 学会安装和使用蒸馏装置。

二、实验原理

当液体物质受热时，其蒸气压随温度升高而增大，待蒸气压增大到与外界施于液面的总压力（通常是大气压力）相等时，液体沸腾，此时的温度称为该液体的沸点。大气压力下将液体加热至沸腾,使之成为蒸气,然后将蒸气冷凝为液体的联合操作称为普通蒸馏。普通蒸馏是分离和纯化液体有机物最常用的方法之一。

利用蒸馏可将沸点相差较大(30 ℃以上)的液态混合物分开。纯液态有机物在蒸馏过程中沸程(沸点范围)较窄(0.5～1 ℃),而混合物的沸程较宽。因此利用蒸馏可以测定沸点,进而判定化合物的纯度,定性地鉴定化合物。应当注意的是,某些有机物往往能和其他组分形成二元或三元恒沸混合物,它们也有固定的沸点,所以不能认为沸点一定的物质都是纯物质。

沸点的测定通常在物质的蒸馏提纯过程中附带进行,称为常量法,此法样品用量较大,一般需 10 mL 以上,其实验装置如图 2-4 所示。

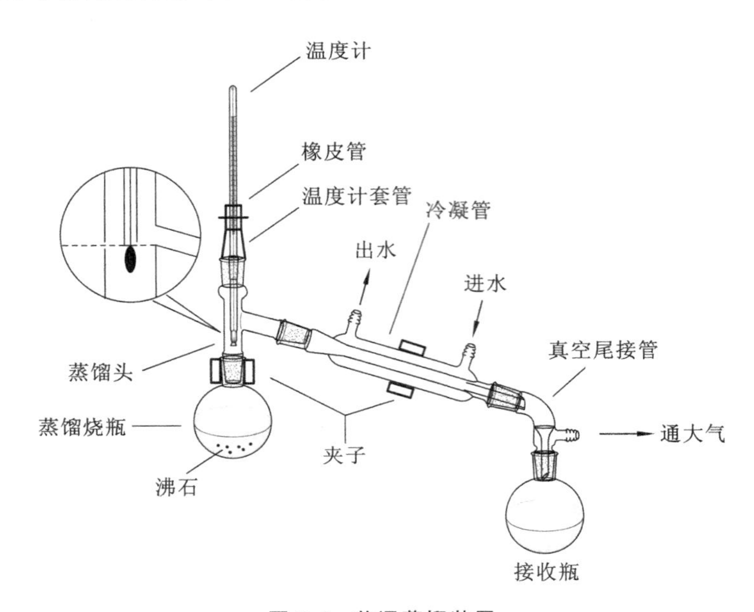

图 2-4　普通蒸馏装置

蒸馏装置主要由蒸馏汽化装置、冷凝装置和接收装置三部分组成。依次包括蒸馏烧瓶、蒸馏头、温度计套管、温度计、冷凝管、尾接管、接收瓶等。

(1) 蒸馏烧瓶。蒸馏烧瓶为蒸馏容器,液体在瓶内受热汽化,蒸气经蒸馏头支管进入冷凝管。蒸馏烧瓶的大小应根据待蒸馏液体的体积决定,通常被蒸馏的液体占蒸馏烧瓶容积的1/3～2/3为宜,否则沸腾时液体易冲出或残留较多。

(2) 温度计。将温度计用橡皮管固定于温度计套管内(或用聚四氟乙烯温度计套塞),然后插入蒸馏头中,调整温度计的位置,使温度计水银球上端恰好与蒸馏头支管下缘在同一水平线上(见图 2-4),以保证在蒸馏时整个水银球能完全被蒸气浸润。

(3) 冷凝管。蒸气在其中冷凝成液体。当待蒸馏液体的沸点低于 140 ℃时,应选用直形冷凝管;当沸点高于 140 ℃时,应选用空气冷凝管。因为温度高时,如用水作为冷却介质,冷凝管内外温差较大,易使冷凝管接口处局部骤冷而断裂。

(4) 接收瓶。常用尾接管与三角烧瓶或圆底烧瓶连接,常压下的蒸馏装置必须通过尾接管的侧管与大气相通,否则当体系受热后液体蒸气压增大时,易发生爆炸引起火灾。如果蒸馏出的物质易受潮分解,可在尾接管上连接一个氯化钙干燥管,以防止湿气的侵入;如果蒸馏出的物质易挥发、易燃或有毒,则需在尾接管上连接气体吸收装置,或通过橡皮管导入下水管道。

三、仪器与试剂

【仪器】圆底烧瓶，蒸馏头，温度计，直形冷凝管，尾接管，三角烧瓶。

【试剂】95％工业乙醇(15 mL)。

【物理常数】

试剂	相对分子质量	熔点/℃	沸点/℃	密度/(g·cm^{-3})	溶解性	n_D^{20}
乙醇	46.07	−114.3	78.4	0.7893	与水混溶	1.3605

四、实验步骤

1. 加料

如图 2-4 所示，按照“由下至上”和“由左至右”的原则安装好仪器，取下温度计和温度计套管，然后将 15 mL 95％工业乙醇通过长颈玻璃漏斗慢慢加入 25 mL 圆底烧瓶中(长颈玻璃漏斗下口处的斜面应低于蒸馏头支管)，投入 2～3 粒沸石[1]，再装上温度计[2]。

2. 加热

加热前，先向冷凝管通入冷却水[3]，将上口流出的水导入水槽中，控制冷却水流速以能保证蒸气充分冷凝为宜，一般只需保持缓慢的水流即可。同时检查装置是否装配严密，以防止在蒸馏过程中蒸气漏出，导致产品损失或发生火灾等事故[4]。

加热时，先小火加热，然后慢慢加大火力，使之沸腾，开始蒸馏，此时应密切注意观察烧瓶内的现象和温度计读数的变化。调节加热套电压，使蒸馏速度以每秒 1～2 滴为宜。记录第一滴馏出液流出时的温度，当温度计读数上升至 77 ℃且稳定不变后[5]，另换一个已称量过的干燥接收瓶收集 77～79 ℃馏分。

3. 停止蒸馏

77～79 ℃馏分蒸完后，如不需要接收第二馏分，可停止蒸馏。注意，即使杂质很少，也不能蒸干，应残留 0.5～1 mL 液体，以保证安全。此时应先停止加热，待稍冷却后馏出物不再流出时，停止通冷却水，拆卸装置，其顺序与安装时相反。

4. 计算回收率

称量所收集馏分并计算回收率。

五、注解和实验指导

【注解】

[1] 沸石的作用是防止液体暴沸，使液体保持平稳沸腾。当液体加热到沸点时，沸石能产生细小的气泡，成为沸腾中心；一旦停止加热后再蒸馏，则原来的沸石失效，应补加新的沸石。若事先忘了加沸石，应立即停止加热，待液体冷却后再补加，否则会引起暴沸，导致液体冲出瓶外，甚至引发火灾。

[2] 若温度计水银球的位置高了，所测沸点会偏低；若位置低了，所测沸点会偏高。

[3] 冷却水从冷凝管支口的下端进，上端出，务必使冷凝管夹层充满冷却水。

[4] 蒸馏沸点较低的易燃液体(如乙醚)时，应用水浴加热，严禁明火加热，且蒸馏速度不能太快，以保证蒸气全部冷凝。如室温较高，接收瓶应置于冷水浴或冰水浴中冷却，并在尾接

管支口处连接一根橡皮管,通入下水管道内或引至室外。

[5] 此时的温度即馏出液的沸点,若所用温度计未经校正,实际测得的沸点可能略有差异。

【预习要求】

(1) 理解沸点和大气压力的关系。

(2) 了解普通蒸馏装置的安装方法及各仪器安装的基本要求和理由。

(3) 思考实验中可能影响沸点测定的因素。

【操作注意事项】

(1) 加热前要检查装置是否严密、正确,要做到准确、端正,各铁夹不要夹得太紧或太松,以免在蒸馏过程中导致玻璃仪器破裂或脱落。

(2) 待蒸馏液体的体积占蒸馏烧瓶体积的 1/3～2/3。

(3) 在加热蒸馏前要加入沸石,因故停止加热后新加沸石,或者补加沸石,都必须待液体冷却后加入,以免发生暴沸。

(4) 控制热浴的温度。热浴的温度应以馏出液的滴出速度为准,一般以比待蒸馏物沸点高出 30 ℃为宜。浴温过高会使蒸馏速度过快,大量蒸气逸出,导致火灾或蒸馏物过热发生分解。

(5) 严禁蒸干,以防意外事故发生。

(6) 注意蒸馏装置的安装、拆卸的顺序。安装时一般从热源处开始,由下至上、由左至右,依次安装。热源应置于升降台上,并保持一定的高度,确保意外发生时能及时撤除热源。拆卸时顺序相反。

六、思考题

(1) 什么叫沸点? 液体的沸点与大气压有何关系?

(2) 当加热后有馏出液流出时,才发现冷凝管未通冷却水,此时能马上通水吗? 如果不行,正确的操作是什么?

(3) 为什么蒸馏系统需要通大气?

(4) 蒸馏时温度计水银球上无液滴意味着什么?

(5) 能否用普通蒸馏的方法将 95%的乙醇进行分离?

Experiment 3 Simple Distillation and Determination of Boiling Point

Ⅰ Objectives

(1) To be familiar with the principle of simple distillation and determination of boiling point.

(2) To know the usage of distillation flask, condenser, etc.

(3) To learn how to set up and use the apparatus for simple distillation.

Ⅱ Principle

When the vapor pressure of the liquid being heated is equal to the external pressure (usually the atmospheric pressure), it boils. The temperature, at that moment, is called boiling point. Should the external pressure vary, so will the boiling point. Simple distillation is a joint operation of heating the boiling liquid to vapor and condensing the vapor into liquid at atmospheric pressure. It is most commonly used in the purification and separation of liquid organic compounds.

Simple distillation is always used to separate liquid mixtures whose boiling points differ by at least 30℃. The boiling range of a pure liquid compound is 0.5～1 ℃, but that of a mixture is wider. Every pure organic compound has a fixed boiling point at a certain pressure. Therefore, simple distillation can be employed to determine the boiling point, then determine the purity of the compound and identify the compound qualitatively. But sometimes an organic compound together with other components may form a binary or ternary azeotrope which also has a definite boiling point. So it is difficult to say whether a liquid is a pure organic compound or not, based on its definite boiling point.

Determination of boiling point is usually included in the process of distillation, which is called regular-amount method. The sample required in this method is generally more than 10 mL. The apparatus for simple distillation is shown in Figure 2-4.

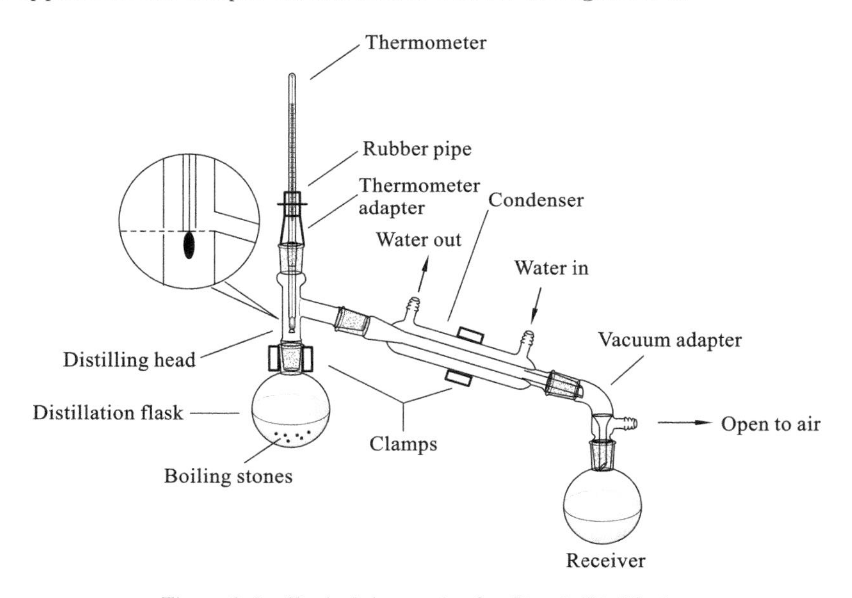

Figure 2-4 Typical Apparatus for Simple Distillation

The apparatus for simple distillation is mainly consist of three parts(i. e. gasification unit, condensing unit and receiving unit), which includes distillation flask, distilling head, thermometer, thermometer adapter, condenser, receiver adapter and receiver.

(1) Distillation flask. Distillation flask is the vessel containing the liquid to be distilled, in which the liquid is heated to vapor. Then the vapor rises through the distilling head into the condenser. The size of the distillation flask is dependent upon the volume of the liquid to be distilled. The volume of the liquid is generally 1/3 to 2/3 of the flask volume, or the boiling liquid may easily rush out.

(2) Thermometer. The thermometer, fixed in a thermometer adapter with a rubber pipe (or a PTFE thermometer sleeve plug), is inserted in the distilling head. It should be positioned so that the top of the mercury bulb is approximately parallel with the bottom of the side-arm outlet (see Figure 2-4), which makes sure that the mercury bulb is completely infiltrated by the vapor.

(3) Condenser. The vapor is condensed into a liquid in the condenser. If the boiling point of the liquid is below 140 ℃, Liebig condenser (i. e. water-cooled condenser) is satisfactory. If the boiling point of the liquid is above 140 ℃, an air condenser should be preferably used to avoid the fracture of condenser resulting from the big temperature difference between inside and outside when using water as a cooling medium.

(4) Receiver. Erlenmeyer flask or round-bottom flask is commonly used as receiver to collect the distillate, which is connected with the condenser using a receiver adapter. The adapter for simple distillation must have a vent to prevent a closed system when heating. The moisture can be excluded by attaching a calcium chloride drying tube to the vent. If the liquid is volatile, flammable or toxic, the vent should be fitted with a piece of rubber tubing, the outlet of which is leaded to a gas absorption device, an open window or drain.

Ⅲ Apparatus and Reagents

【Apparatus】 round-bottom flask, distilling head, thermometer, Liebig condenser, receiver adapter, and Erlenmeyer flask.

【Reagents】 95% industrial alcohol (15 mL).

【Physical constants】

Reagent	M_w	m. p. /℃	b. p. /℃	ρ/(g · cm^{-3})	Solubility	n_D^{20}
Ethanol	46.07	−114.3	78.4	0.7893	Miscible with water	1.3605

Ⅳ Procedures

1. Charging

Assemble a simple distillation apparatus shown in Figure 2-4 according to the "bottom-to-top" and "left-to-right" principle. Take down the thermometer and its adapter, and then add 15 mL of 95% industrial alcohol into a 25 mL round-bottom flask through a long-stem funnel, of which the bottom should be lower than the side-arm of the distilling head. Place several boiling stones in the flask[1], and reassemble the thermometer[2].

2. Heating

Before heating, introduce the cooling water into the condenser[3], and adjust the water flow to a modest flow rate. No benefit is gained from a fast flow. Too much water pressure will cause the tubing to pop off. Also check whether each ground joint is firmly fitted with the other without twisting or loosing to prevent the leakage of vapor, which resulting in loss of the product or fire[4].

First heat slowly with a heating mantle, and then gradually increase the power, make the liquid boil, and start distilling. Great attention should be paid to the phenomenon in the flask and change of the temperature. Adjust the heating power, maintain the distillation at a rate of one or two drops per second, and record the temperature of first drop of the distillate. When the temperature rises to 77 ℃ and remains stable[5], replace the receiver with a weighed and dry flask to collect the distillate in the range of 77～79 ℃.

3. Ceasing of the distillation

If no more distillates to be collected, cease the distillation. Great care should be taken not to distill to dryness because, in some cases, high-boiling explosive peroxides can become concentrated. Small volume of the liquid (usually 0.5～1 mL) should be remained in the flask. Take off the heating mantle, and cease the cooling water after no distillate comes out. Disassemble the apparatus following the "right-to-left" and "top-to-bottom" principle.

4. Calculating the percentage yield

Weigh the collected distillate and calculate the percentage yield.

Ⅴ Notes and Instructions

【Notes】

[1] Adding boiling stones can promote even boiling to prevent bumping. The boiling stones, absorbing large quantities of gases, will release small bubbles to be boiling centers when boiling. Once ceasing the heating, the distillation cannot be restarted unless the fresh ones are added. They should never be added to the hot liquid.

[2] If the thermometer is highly positioned, the boiling point measured will be lower than the theoretical value; if the thermometer is lowly positioned, the boiling point measured will be higher than the theoretical value.

[3] Note the location of the "water in" and "water out" nipples on the condenser. The tubing carrying the incoming water is always attached to the lower point, which ensures that the condenser is filled with the cooling water at all times.

[4] When the flammable liquid with low boiling point (e. g. ether) is distilled, it should not be heated by open flame, but a water bath. The distillation rate should not be too fast, ensuring the complete condensation of the vapor. If the room temperature is high, the receiver should be cooled in a cold-water bath or ice-water bath, and a piece of rubber tubing be attached to the vent of the adapter. The outlet of the tubing is leaded to an open window or drain.

[5] At this moment, the temperature is the boiling point of the distillate. If the uncorrected thermometers are used in the experiment, the readings may be slightly different.

【Requirements for preview】

(1) To understand the relationship between boiling point and atmospheric pressure.

(2) To know how to set up the apparatus for simple distillation, and basic requirements for each part of the apparatus.

(3) To know what factors the boiling point measured is governed by.

【Experimental precautions】

(1) Before heating, check whether the apparatus is firmly and correctly fitted. It should be regular in both horizontal and vertical directions. Just tighten the clamps only enough to hold each part in place. Align the jaws of the clamp parallel to the piece of glassware being clamped. This enables the clamp to be tightened without increasing of stressing to break the glassware or pulling another joint loose.

(2) The volume of the liquid to be distilled is generally 1/3 to 2/3 of the flask volume.

(3) Never add the boiling stones into a hot liquid. The fresh ones cannot be added unless the liquid cools down.

(4) Satisfactory heating bath is selected according to the boiling point of the liquid, of which the heating temperature is preferably higher by a magnitude of 30 ℃ than the boiling point of the liquid. Excessive heating may cause too-rapid distillation and leakage of large quantity of vapor, resulting in fire or decomposition of the distillates.

(5) Never distill to dryness. The residues may ignite or explode with great violence.

(6) Note the "bottom-to-top" and "left-to-right" principle for assembling and the reverse one for disassembling. The distillation flask, the starting point of the whole apparatus, should be elevated at a certain height (usually 15 cm or so) above the bench, which allows the quick removal of heating source when the accident happens.

Ⅵ Post-lab Questions

(1) What is the boiling point? How about the relationship between boiling point and atmospheric pressure?

(2) When the distillate comes out, the operator finds that there is no cooling water in the condenser. Can the cooling water be introduced into the condenser immediately? If not, what is the correct operation?

(3) Why should the distillation system have an opening to the air at the end?

(4) There are no droplets on the mercury bulb of the thermometer during the distillation. What does this phenomenon mean?

(5) Can the 95% ethanol be separated by simple distillation?

Ⅶ Verbs

simple distillation 普通蒸馏;
distillation apparatus 蒸馏装置;
vapor pressure 蒸气压;
boiling stone 沸石;
boiling range 沸程;
distillate 馏分;
azeotrope 恒沸物;
regular-amount method 常量法

实验四 重 结 晶

一、实验目的

(1) 了解重结晶的基本原理。

(2) 初步学会用重结晶的方法提纯固体有机化合物。

(3) 掌握热过滤和抽滤等操作。

二、实验原理

重结晶是纯化固体有机化合物的重要方法之一,其原理是利用被提纯物质与杂质在某种溶剂中溶解度的不同而达到分离纯化的目的。必须注意,杂质含量过高对重结晶极为不利,会影响结晶速率,甚至妨碍结晶的生成。重结晶一般适用于杂质含量低于5%的固体化合物,所以将反应粗产物直接重结晶是不适宜的,应采用其他方法进行初步提纯,如萃取、水蒸气蒸馏等,然后进行重结晶。

重结晶操作的一般过程如下:选择溶剂→溶解固体(制备热的近饱和溶液)→脱色→除去不溶性杂质→晶体的析出→晶体的收集与洗涤→晶体的干燥。

1. 溶剂的选择

正确选择溶剂,对重结晶具有重要意义。所选择溶剂必须符合下列条件:

(1) 不与重结晶的物质发生化学反应;

(2) 在高温时,重结晶物质在溶剂中溶解度较大,而在低温时则很小;

(3) 杂质的溶解度应很大(使杂质留在母液中不随被提纯物的晶体析出,以便分离)或很小(在制成热的近饱和溶液后,趁热过滤除去杂质);

(4) 能得到较好的晶体。

此外,沸点适中、价廉易得、毒性低、回收率高、操作安全的溶剂更佳。

在选择溶剂时,若根据化学手册或文献查不到合适的溶剂,则应根据"相似相溶"原理,通过实验来确定。其方法如下:取几只小试管,各放入约0.2 g待重结晶的固体样品,分别加入0.5~1 mL不同种类的溶剂,加热至完全溶解,冷却后能析出最多晶体的溶剂,一般认为是最合适的溶剂。如果固体物质在3 mL热溶剂中仍不能全部溶解,则该溶剂不适用于重结晶;如果固体在热溶剂中能溶解,而冷却后无晶体析出,可用玻璃棒在液面下的试管内壁上摩擦,以促使晶体析出,若仍未析出晶体,则说明待重结晶物质在该溶剂中的溶解度很大,也不适于重结晶。

如果固体物质易溶于某一溶剂(称为良溶剂)而难溶于另一溶剂(称为不良溶剂),而这两种溶剂能互溶,可将两者配成一定比例的混合溶剂来进行实验。常用的混合溶剂体系有乙醇-水、乙醇-乙醚、乙醇-丙酮、乙醚-石油醚、苯-石油醚等。

重结晶操作中常用的溶剂以极性递减的顺序列于表2-3中,表中相邻的溶剂由于极性相似,可以互溶,每一种溶剂一般可溶解与其化学结构相似的化合物。

表 2-3 常用的重结晶溶剂

溶剂	沸点/℃	备注
水	100	最优先选择的溶剂,但由于沸点及汽化热较高,较难干燥
冰乙酸	118	不易燃烧,蒸气有刺激性气味
二甲亚砜	189	不常用,沸点高,难以从晶体中除去
甲醇	64.5	易燃,有毒
95%乙醇	78	易燃,最常用的重结晶溶剂之一
丙酮	56.2	易燃,但其沸点较低,和室温差异偏小
乙酸乙酯	77.2	易燃,较高沸点和挥发性的完美结合,最常用溶剂之一
二氯甲烷	39.8	难燃,有毒,但沸点太低
三氯甲烷	61.3	难燃,蒸气有毒
乙醚	34.6	易燃,应尽量避免使用
苯	80.1	易燃,蒸气剧毒
甲苯	110.6	易燃,蒸气毒性低于苯,为苯的最佳替代溶剂
四氯化碳	76.8	难燃,蒸气有毒
环己烷	80.7	易燃,常用于重结晶非极性化合物
石油醚	30～60	易燃,短链烷烃混合物,其主要成分为正戊烷
	60～90	易燃,短链烷烃混合物,其性质类似于正己烷和环己烷

2. 固体溶解

通常在锥形瓶或烧杯中进行重结晶,但使用易挥发或易燃的有机溶剂时,为了避免溶剂的挥发而引发火灾,则需将待重结晶物质放入圆底烧瓶中,圆底烧瓶上应装有回流冷凝管,溶剂可由回流冷凝管上口加入。先加入少量溶剂,加热至沸腾,然后逐渐地添加溶剂(加入后再加热煮沸),直到固体全部溶解为止。

应注意,不因待重结晶的物质中含有不溶解的杂质而加入过量的溶剂。随时观察固体的溶解情况,当怀疑未溶解固体为不溶性杂质时,停止加入溶剂。加热至沸腾,如果未溶解固体量不发生变化,则利用热过滤除去不溶性杂质;如果溶液有颜色,则需先进行脱色;如果制得的溶液无不溶性杂质且没有颜色,可直接冷却结晶。

为防止在热过滤时由于溶剂的挥发和温度的下降导致溶解度降低而较快地在滤纸上或漏斗中结晶析出,造成产品损失,一般需补加少量溶剂(理论量的10%左右),将饱和溶液略微稀释。除高沸点溶剂外,一般在水浴上加热;在加入可燃性溶剂时,应先熄灭明火。

3. 脱色

若所得溶液中存在有色物质,一般可用活性炭脱色。活性炭的用量,以能完全除去颜色为度,为了避免过量,应分成小量,分批加入,用量一般为固体样品的1%～5%。必须待热的饱和溶液稍冷后再加入适量活性炭,在不断搅拌下煮沸5～10 min,然后趁热过滤。严禁向正在沸腾的溶液中加入活性炭,以免溶液暴沸而溅出,造成产品损失,甚至烫伤。

4. 不溶性杂质的除去

如果所得到的热饱和溶液中含有不溶解的杂质,应趁热过滤除去。热过滤有两种方法:常压热过滤和减压热过滤(抽滤)。

常压热过滤操作应选用短颈径粗的玻璃漏斗,使用折叠式滤纸(折叠方法见图 2-5)和保温漏斗(见图 2-6),以便过滤尽快完成。

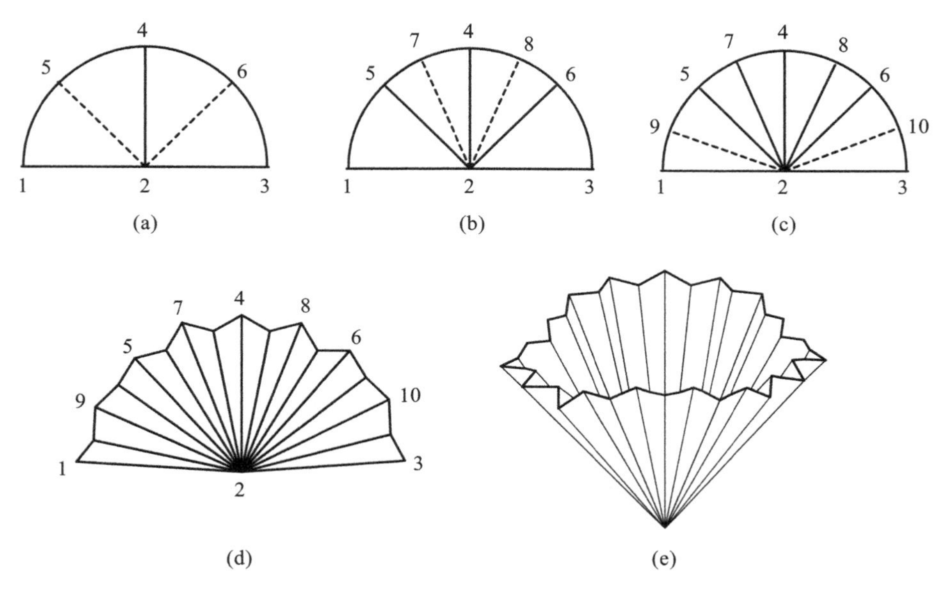

图 2-5　折叠式滤纸的折法

折叠式滤纸具有较大的有效面积,因而过滤速率更快。每次折叠时,在折纹的集中点切勿重压,以免过滤时滤纸中央破裂。

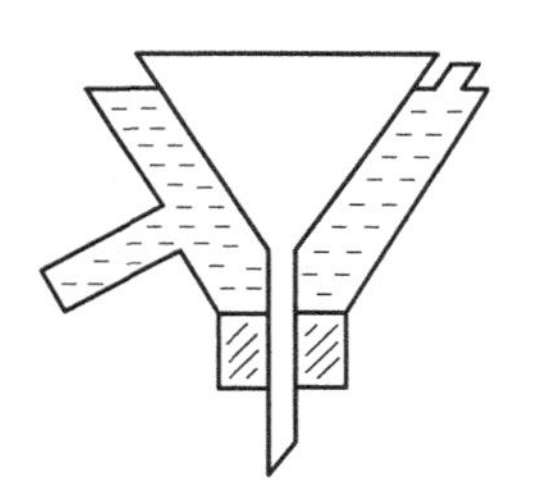

图 2-6　保温漏斗

保温漏斗具有铜制的外壳,里面插一个短颈的三角漏斗,在外壳的支管处加热,即可把夹层中的水烧热而使漏斗保温。如果没有保温漏斗,也可用短颈三角漏斗代替,预热后趁热过滤。操作时,一般需准备三个容器,置于水浴或加热板上加热。其中一个盛有待过滤溶液;一个盛有适量的水和漏斗,以便预热漏斗(也可置于烘箱中预热);第三个盛有适量重结晶溶剂,以备洗涤之用。

过滤时,先用少量热溶剂润湿滤纸,用毛巾包住容器,然后将待过滤的溶液沿玻璃棒小心倒入漏斗中的折叠式滤纸内,并用表面皿盖住漏斗,以减少溶剂的挥发。剩余的溶液应继续加热,以避免冷却结晶。过滤完毕后,可用少量热溶剂冲洗滤纸,将滤纸上析出的晶体重新溶解。

减压热过滤的装置如图 2-7 所示,减压过滤时,为了避免漏斗破裂及在漏斗中析出晶体,应事先将布氏漏斗和抽滤瓶用热水浴、水蒸气浴或在电烘箱中预热,然后趁热过滤。过滤时动作要快,以免由于温度降低而在滤纸上析出结晶。减压热过滤的优点是过滤速度快,缺点是当溶剂的沸点较低时,减压会使热溶剂蒸发或沸腾,导致溶液浓度变大,使晶体过早析出。

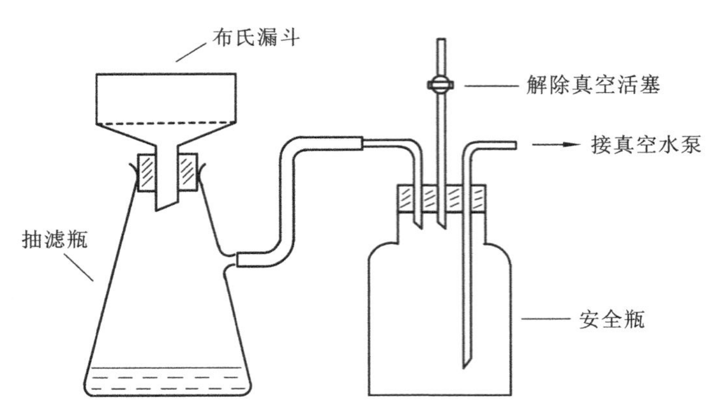

图 2-7　减压热过滤装置

5. 晶体的析出

热过滤后的溶液冷却后,晶体即析出,但是冷却条件不同,晶体析出的情况也有所不同。一般来说,溶液浓度较大、冷却速度较快时,析出的晶体较细,也不够纯净。由于热的溶液在碰到冷的抽滤瓶时,往往很快析出晶体,但晶体质量较差,因此常将溶液重新加热使晶体完全溶解,然后将溶液静置,使其自然、缓慢地冷却,不要骤冷和剧烈搅拌,由此可得颗粒大、形状好、纯度高的晶体。室温下冷却结晶结束后,可将烧瓶置于冰-水浴中进一步冷却,以得到更多的晶体。

如果溶液冷却后仍不结晶,可用玻璃棒摩擦容器内壁引发晶体形成,或投入"晶种"(同一物质的晶体),以供给晶核,促使晶体较快地形成。晶种加入量不宜过多,而且加入后不要搅拌,以免晶体析出太快,影响产品的纯度。

如果待重结晶的物质不析出晶体而析出油状物,其原因之一是热的饱和溶液的温度高于或接近待重结晶物质的熔点。可重新加热溶液至澄清后,使其自然冷却,当有油状物出现时,立即剧烈搅拌,使油状物在均匀分散的条件下固化,也可搅拌至油状物完全消失,冷却后即析出晶体。

6. 晶体的收集与洗涤

晶体析出后,常采用抽滤(减压过滤)的方法使留在溶剂中的可溶性杂质与晶体彻底分离,其装置同减压热过滤装置(见图 2-7)。在布氏漏斗中铺一张比漏斗内径略小的圆形滤纸(以能恰好盖住所有的小孔为宜),过滤前应先用溶剂润湿滤纸,打开水泵,关闭安全瓶活塞,抽气,使滤纸紧贴在漏斗上。小心地将要过滤的混合物倒入漏斗中,使固体均匀地分布在整个滤纸面上,并用少量滤液将黏附在容器壁上的晶体洗出,合并入漏斗中,继续抽气,同时用空心玻璃塞挤压晶体,直至布氏漏斗下端不再滴出溶剂为止。缓慢旋开安全瓶上活塞,使其与大气相通,再关闭水泵。

滤得的固体(俗称滤饼)表面有母液残留,需用干净溶剂加以洗涤。将少量溶剂均匀地洒在滤饼上,并用玻璃棒或刮刀小心翻动晶体(不能使滤纸松动),使全部晶体刚好被溶剂浸润,待有滤液从漏斗下端滴下时,重新抽气,除去溶剂,重复操作两次,即可将滤饼洗净。

若使用的重结晶溶剂的沸点较高,因其较难挥发,在用原溶剂洗涤一次后,可用低沸点的溶剂再洗一次,以利于晶体的干燥(注意:此溶剂必须能与原来的溶剂相混溶,且对晶体是微溶或不溶的)。

7. 晶体的干燥

用重结晶法纯化后的晶体，其表面还吸附有少量溶剂，为了保证产品的纯度，应根据所用溶剂及晶体的性质选择适当的方法进行干燥，见实验二。

三、仪器与试剂

【仪器】烧杯，三角漏斗，布氏漏斗，抽滤瓶，圆底烧瓶，冷凝管。

【试剂】粗苯甲酸，粗乙酰苯胺，粗乙酰水杨酸。

【物理常数】

试　剂	相对分子质量	熔点/℃	沸点/℃	密度/($g \cdot cm^{-3}$)	溶解性(25 ℃)
苯甲酸	122.12	122.4	249	1.2659	微溶于水
乙酰苯胺	135.17	114.3	304	1.2190	微溶于水
乙酰水杨酸	180.16	135.0	—	1.35	微溶于水，易溶于乙醇

四、实验步骤

1. 用水重结晶苯甲酸[1]

(1) 称取 0.5 g 粗苯甲酸，置于 25 mL 烧杯中，再加入 10 mL 水和几粒沸石。加热至沸，并用玻璃棒不断搅动，使固体溶解。若有少量未溶解的固体，可继续加入少量热水至固体全部溶解[2]。

(2) 若溶液颜色较深，则移去热源，稍冷后加入适量活性炭，继续煮沸 5～10 min。

(3) 热过滤以除去溶液中的不溶性杂质。

(4) 将滤液静置冷却后，即有苯甲酸晶体析出。用布氏漏斗抽滤使晶体和母液分离，并用空心塞挤压固体，将母液尽量除去。停止抽气后，用少量干净的冷水洗涤晶体 1～2 次，并重新抽干。

(5) 将晶体转移至表面皿上，置于红外灯下烘干[3]。

(6) 干燥完全后，测定熔点，并与粗产物的熔点进行比较，称重并计算回收率。

2. 用水重结晶乙酰苯胺[4]

(1) 称取 0.5 g 粗乙酰苯胺，置于 25 mL 烧杯中，再加入适量的水[5]和几粒沸石。加热至沸，使固体全部溶解[6]。

(2) 以下步骤与要求和苯甲酸重结晶相同。

3. 用乙醇-水混合溶剂重结晶乙酰水杨酸

(1) 在装有回流冷凝管的 25 mL 圆底烧瓶中，加入 0.5 g 粗乙酰水杨酸、5～6 mL 95%的乙醇和几粒沸石。接通冷凝水后加热至沸，并不时摇动以加速溶解。

(2) 待固体完全溶解后，趁热滤入另一圆底烧瓶中[7]，然后加装冷凝管，继续加热至沸，同时用滴管自冷凝管上端滴加蒸馏水，至沸腾条件下出现混浊而不再变澄清为止。

(3) 溶液于室温下放置，自然冷却结晶。

(4) 以下步骤与要求和苯甲酸重结晶相同。

五、注解和实验指导

【注解】

[1] 苯甲酸在水中的溶解度为:0.21 g(17.5 ℃)、0.35 g(25 ℃)、2.2 g(75 ℃)、2.7 g(80 ℃)、5.9 g(100 ℃)。

[2] 每次加入 1～2 mL 热水,若加入溶剂且加热后并未使不溶物减少,则可能是不溶性杂质,可不必再加溶剂。

[3] 注意不要使温度超过 100 ℃。

[4] 乙酰苯胺在水中的溶解度为:0.46 g(20 ℃)、0.56 g(25 ℃)、0.84 g(50 ℃)、3.45 g(80 ℃)、5.5 g(100 ℃)。

[5] 0.5 g 粗乙酰苯胺需加水 15～20 mL。

[6] 溶解过程中会出现油状物,此油状物不是杂质,而是乙酰苯胺和水形成的共熔物,此时可继续加热或加水至油状物全部溶解。

[7] 乙醇为易燃溶剂,过滤时周围切不可有明火。

【预习要求】

(1) 了解重结晶的基本原理和一般过程。

(2) 理解重结晶中混合溶剂的选择要求和操作与单一溶剂的不同。

(3) 熟悉热过滤和抽滤的装置及操作要求。

【操作注意事项】

(1) 若所用溶剂为有机溶剂,样品的溶解需在回流装置中进行,补加溶剂及热过滤时,必须先熄灭周围明火。

(2) 用活性炭脱色时,切不可将活性炭投入正在沸腾的溶液中。

(3) 热过滤时,操作要迅速,使溶液尽快通过漏斗,以防止温度下降使晶体在漏斗中析出。

(4) 停止抽滤时,应先将抽滤瓶与真空水泵间的橡皮管断开,或者将安全瓶上的活塞打开与大气相通,再关闭水泵,以防止水倒吸入抽滤瓶内。

六、思考题

(1) 重结晶操作包括哪几个步骤?各步骤的目的分别是什么?

(2) 为什么重结晶时溶剂的用量不能过多也不能过少?应如何确定正确的溶剂用量?

(3) 用有机溶剂重结晶时,哪些操作易导致火灾?正确操作是什么?

(4) 如何选择合适的重结晶溶剂?

Experiment 4　Recrystallization

Ⅰ Objectives

(1) To know the basic principle of recrystallization.

(2) To learn how to purify the solid organic compounds by recrystallization.

(3) To master the operation of hot filtration and vacuum filtration.

Ⅱ Principle

Recrystallization is the most important method for the purification of solid organic compounds from impurities. This method is based upon the significant differences in their solubility in a given solvent or mixture of solvents. It should be noted that the high impurity content is extremely detrimental to the crystallization, which may decrease the crystallization rate, even interfere with the formation of the crystals. Therefore, recrystallization is only employed to purify the solid compounds with low impurity content(usually less than 5%). It is not satisfactory to the direct recrystallization of the raw products obtained from a reaction, unless they are preliminarily purified by extraction or steam distillation.

The general process of recrystallization can be broken into seven discrete steps: Choosing the solvent, dissolving the solute(i. e. preparation of a nearly saturated solution at the boiling point), decolorizing the solution, removing suspended solids, crystallizing the solute, collecting and washing the crystals, and drying the product.

1. Choosing the solvent

The most desirable characteristics of a solvent for recrystallization are as follows:

(1) It will not react chemically with the solute to be purified.

(2) It will dissolve the solute when the solution is hot but not when the solution is cold.

(3) It will not dissolve the impurities at all(so they will be removed from the nearly saturated solution by hot filtration), or it will dissolve them very well (so they won't crystallize out along with the solute).

(4) It will yield well-formed crystals of the solute.

If two or more solvents appear to be equally suitable for recrystallization, the final selection will depend upon such factors as ease of manipulation, toxicity, flammability and cost.

If no information is already available in chemistry handbook or literature, it should be chosen by experimentation according to the principle of "like dissolves like". The procedure of picking a solvent is described as follows: Put a small amount of the impure solute in several small test tubes(usually about 0.2 g for each tube), and then add 0.5～1 mL of different solvents into the tubes, respectively. Heat the solution until all the solute dissolves, and then cool down to room temperature. The one, in which the maximum crystals are obtained, is generally considered to be the most desirable solvent. If the solute is still not completely dissolved in 3 mL of hot solvent, this solvent is unsuitable. If the solute dissolves in the hot solvent, but no crystallization occurs when the solution is cooled, even after scratching the inwall of the tube below the surface of the solution with a glass rod, this indicates that the solute to be purified has large solubility in the solvent. It should be rejected.

If the solute is found to be far too soluble in one solvent(called as the better solvent) and much too insoluble in another solvent (called as the poorer solvent), mixed solvents or "solvent pairs" may frequently be used for recrystallization with excellent results. The two solvents must, of course, be completely miscible. Solvent pairs which may often be employed include: Ethanol and water, ethanol and diethyl ether, ethanol and acetone, diethyl ether and petroleum ether, benzene and petroleum ether, etc.

Some common solvents available for recrystallization and their properties are presented in Table 2-3, broadly in the order of decreasing polarity of the solvent. Solvents adjacent to each other in the list will dissolve in each other, i. e. , they are miscible with each other, and each solvent will, in general, dissolve substances that are similar to it in chemical structure.

Table 2-3　Common Solvents for Recrystallization

Solvent	b. p. /℃	Remark
Water(distilled)	100	To be used whenever suitable, difficult to be removed
Acetic acid(glacial)	118	Not very flammable, pungent vapors
Dimethyl sulfoxide	189	Difficult to be removed
Methanol	64.5	Flammable, toxic
95% Ethanol	78	Flammable, most commonly used
Acetone	56.2	Flammable, lower boiling point
Ethyl acetate	77.2	Flammable, excellent combination of higher b. p. and volatility
Dichloromethane	39.8	Non-flammable, toxic, not a good recrystallization solvent
Chloroform	61.3	Non-flammable, vapor toxic
Diethyl ether	34.6	Highly flammable, avoid whenever possible
Benzene	80.1	Flammable, vapor highly toxic
Toluene	110.6	Flammable, in place of benzene whenever possible
Carbon tetrachloride	76.8	Non-flammable, vapor toxic
Cyclohexane	80.7	Flammable, usually used to recrystallize nonpolar compounds
Petroleum ether	30～60	Flammable, mixture of hydrocarbons, mainly pentane
	60～90	Flammable, mixture of hydrocarbons, similar to hexane

2. Dissolving the solute

Recrystallization is generally carried out in an Erlenmeyer flask or a beaker. While a round-bottom flask of suitable size fitted with a reflux condenser must be employed, when using organic solvent with high flammability and volatility. Place the substance to be recrystallized in an appropriate flask, add enough solvent(less than the required quantity) to cover the solute and several boiling stones, and then heat the flask on a water bath(if the solvent boils below 90 ℃) or a heating mantle until the solvent boils. Stir the mixture with a glass rod, better, swirl it to promote dissolution. Add solvent gradually, keeping the mixture at the boiling, until all of the solute dissolves.

Be careful not to add too much solvent. Note how rapidly most of the solute dissolves and then stop adding solvent when you suspect that the undissolved solid is the insoluble impurity. Allow the solvent to boil, and if no further solute dissolves, proceed to Step 4 to remove suspended solids from the solution by hot filtration, or if the solution is colored, go to Step 3 to carry out the decolorization process. If the solution is clear, proceed directly to Step 5, crystallizing the solute.

In order to prevent the solute from crystallizing quickly on the filter paper or in the funnel, which resulting from the slightly evaporation of the solvent and decreasing of the temperature, a little more solvent (generally about 10% of the required quantity) should be added to transform the saturated solution to a slightly dilute one. With the exception of the solvents with high boiling point, the others are usually heated on a water bath to dissolve the solute. When adding the flammable solvent, the open flame must be extinguished.

3. Decolorizing the solution

Occasionally an organic reaction will produce high molecular weight by-products that are highly colored. These colored impurities can be adsorbed onto the surface of activated charcoal, which has an extremely large surface area (several hundred square meters) per gram. If too little charcoal is added, the solution will still be colored after filtration, making repetition necessary; if too much is added, it will absorb some of the product in addition to the impurities. Gradually add a small amount (1% ~ 5% of the solute weight is sufficient) of activated charcoal to the colored solution and then boil the solution for 5~10 min. Be careful not to add the charcoal pieces to a superheated solution; the charcoal functions like hundreds of boiling stones and will cause the solution to boil over. Remove the activated charcoal by filtration as described in Step 4.

4. Removing suspended solids

The insoluble impurities or the activated charcoal must be removed from the boiling or hot solution before undue cooling has occurred. This process, called as hot filtration, commonly includes gravity filtration and vacuum filtration.

The most common method for the removal of suspended solids is gravity filtration through a fluted filter paper (see Figure 2-5) supported by a heat preservation funnel (see Figure 2-6), which makes the filtration complete as soon as possible.

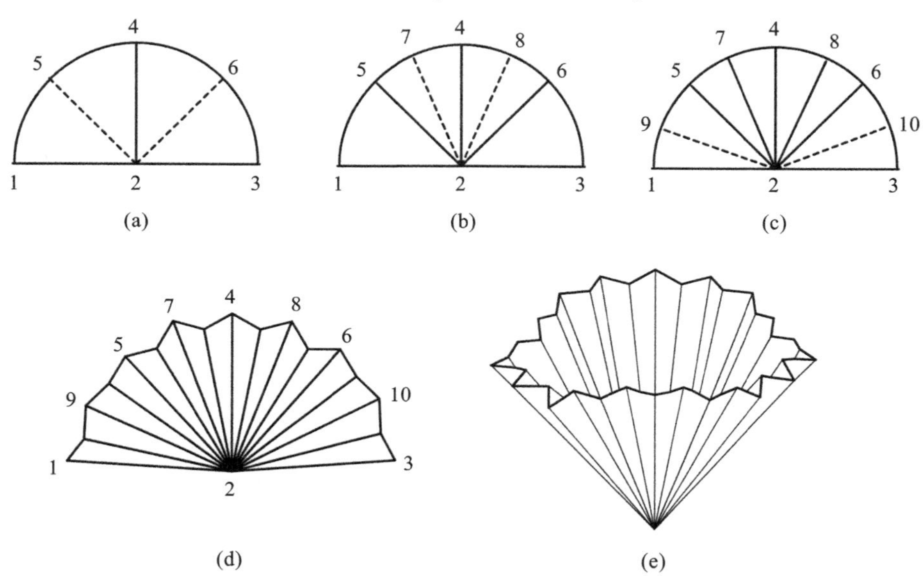

Figure 2-5 Preparation of a Fluted Filter Paper

The fluted filter paper has a larger effective area, and thus a faster filtration rate. As shown in Figure 2-5, the filter paper is first folded in half and again in quarters to produce the folding lines 1-2,2-3 and 2-4;the line 1-2 is then folded onto 2-4 and the line 2-3 onto 2-4, producing,when the paper is opened,new folding lines 2-5 and 2-6(see Figure 2-5(a)). The folding fifty-fifty between 2-4 and 2-5 produces the new line 2-7,and then continued to give 2-8,2-9 and 2-10, respectively(see Figure 2-5(b)and(c)). The final operation consists in making a fold in each of the eight segments(between 2-3 and 2-10, between 2-10 and 2-6, etc.)in a direction opposite to the first series of folds(i. e. the folds are made outwards instead of inwards as at first). The result is a fan arrangement,and upon opening,the fluted filter paper is obtained(see Figure 2-5(d)and(e)). In each operation,the focal point of the folds should not be heavily stressed to avoid the fracture of the filter paper in central part.

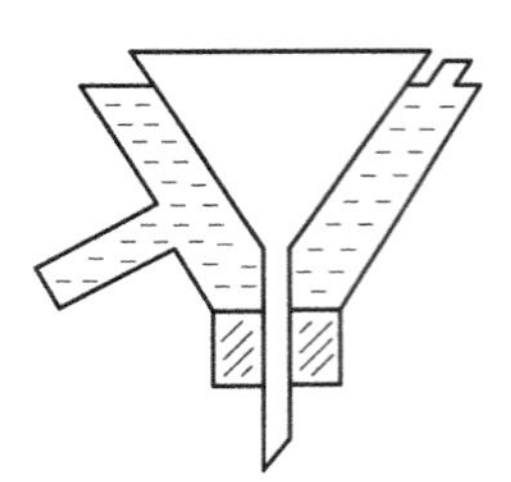

Figure 2-6　Heat Preservation Funnel

As shown in Figure 2-6, a short-stem funnel is inserted in a jacket made of copper. Heating the side arm of the jacket, in which the water will be heated to keep the funnel warm. If the heat preservation funnel is not available in the lab,a preheated short-stem funnel can also be used to replace it. Note that the preheating should be immediately followed by the filtration. In general, three vessels on a water bath or hot plate will be needed for this process—one to contain the solution to be filtered,one to contain a few milliliters of water and a short-stem funnel for the purpose of preheating(the funnel can also be preheated in an oven),and the third to contain several milliliters of the crystallizing solvent to be used for rinsing purposes.

First wet the filter paper with a small amount of hot solvent(from the third vessel). Grasp the vessel in a towel,and then pour the boiling solution at a steady rate into the paper, guided by a glass rod. Cover the funnel with a watch glass to decrease the evaporation of the hot solvent. The remaining solution should continue to be heated to avoid crystallization. Finally,check to see whether crystallization is occurring on the filter paper or in the funnel. If it does,add a small amount of boiling solvent until the crystals dissolve.

The apparatus for removing the suspended solids from the boiling solution by filtration under reduced pressure is shown in Figure 2-7. It is just the same as the apparatus for vacuum filtration,of which the usage in detail will be described in Step 6. The Büchner funnel used in this apparatus should also be preheated on a hot-water bath, steam bath or in an oven, immediately followed by the filtration. The filtration rate is much higher for the vacuum filtration than for the gravity filtration. But the solvent with relatively low boiling point may more easily evaporate under reduced pressure, or even boil in the filter flask at low temperature,thus resulting in crystallizing prematurely.

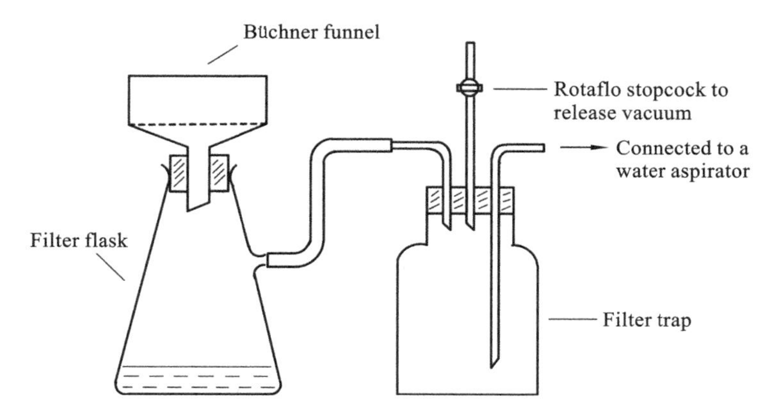

Figure 2-7　Apparatus for Vacuum Filtration

5. Crystallizing the solute

The saturated solution, obtained by hot filtration, is allowed to cool to room temperature. Crystallization should begin immediately. The quality of the crystals will be different under different cooling conditions. If the concentration of the solution is large and the cooling rate is high, the crystals obtained will be much smaller, and also not pure enough. If the crystallization has occurred in the process of hot filtration, the solution should be reheated to dissolve the crystals, and then stood on the bench to cool slowly without being disturbed, thus giving the large, well-formed and highly pure crystals. Once the crystallization ceases at room temperature, the flask should be placed in an ice-water bath to cool further.

If no crystallization occurs after cooling, scratch the inside of the vessel with a glass rod or add a seed crystal(i. e. crystals of the same substance) to provide nucleation center, thus promoting the formation of the crystals. Do not add too much seed crystals and stir the solution, otherwise the crystals will precipitate quickly with low purity.

The separation of a second liquid phase, commonly known as "oil", instead of the expected crystalline solid, sometimes occurs during recrystallization. One of the reasons is that the temperature of the hot solution is higher than or close to the melting point of the substance to be recrystallized. It is probably best to reheat the mixture until a clear solution is obtained, and allow it to cool spontaneously; immediately the oil commences to separate, the mixture is vigorously stirred so that the oil is well dispersed in the solution. Eventually, the crystals will separate and grow in the bulk of the solution. When all the oil has disappeared, stirring may be stopped and the crystals allowed to accumulate.

6. Collecting and washing the crystals

Once the crystallization is complete, the apparatus for vacuum filtration is generally employed to separate the crystals from the mother liquor containing soluble impurities. As shown in Figure 2-7, a porcelain Büchner funnel(the bevel of the neck of the funnel should face the side arm of the filter flask)is fitted with a rubber stopper, which is seated in the neck of a filter flask; the side arm of the filter flask is connected to a filter trap using heavy-wall rubber tubing, then to a water aspirator; the filter trap is fitted with a two-way Rotaflo stopcock to regulate the pressure in the apparatus, which is essential since a sudden fall in water pressure may result in the water being sucked back and contaminating the filtrate.

A filter paper is selected of such size that it covers the entire perforated plate, but its diameter should be slightly less than the inside diameter of the funnel. The filter paper is moistened with a few drops of cold solvent and the suction of the water aspirator applied, then the filter paper should adhere firmly to, and completely cover the perforated plate of the funnel and thus prevent any crystals escaping around the edge. Pour the slurry of crystals into the funnel carefully, and transfer the remaining crystals by rinsing with a small amount of the filtrate (not the fresh solvent). When taking the filtrate out of the filter flask, the latter must be disconnected from the aspirator to avoid the suck-back. The suction is continued until most of the solvent has passed through and this is facilitated by pressing the solid down with a glass stopper to leave a uniformly flat, pressed surface (i. e. filter cake). Open the stopcock on the trap slowly, and then turn off the water aspirator.

The filter cake is then washed with fresh and cold solvent to remove the mother liquor. Add a small amount of the solvent on the filter cake and loosen the crystals carefully with a glass rod or scraper (do not loosen the filter paper). Be sure that all the crystals are exactly immersed in the solvent, and then suck again to remove the solvent. Repeat this process for two times, thus giving a filter cake without mother liquor.

If the recrystallization solvent with high boiling point is used, it is difficult to dry the crystals after washing. It is probably best to wash the crystals again with a low-boiling-point solvent just after the original one (note that this solvent must be miscible with the original one and cannot dissolve the crystals).

7. Drying the product

Once the crystals have been washed on the Büchner funnel, press them down with a clean glass stopper and allow air to pass through them under suction until they are substantially dry. According to the properties of the solvent and crystals, final drying can be carried out using the drying methods described in Experiment 2.

Ⅲ Apparatus and Reagents

【Apparatus】 Beaker, short-stem funnel, Büchner funnel, filter flask, round-bottom flask, filter paper, infrared lamp, condenser.

【Reagents】 Impure benzoic acid (0.5 g), impure acetylaniline (0.5 g), impure acetylsalicylic acid (0.5 g).

【Physical constants】

Reagent	M_w	m. p. /℃	b. p. /℃	$\rho/(g \cdot cm^{-3})$	Water Solubility
Benzoic acid	122.12	122.4	249	1.2659	Slightly soluble
Acetylaniline	135.17	114.3	304	1.2190	Slightly soluble
Acetylsalicylic acid	180.16	135.0	—	1.35	Slightly soluble

Ⅳ Procedures

1. Recrystallization of benzoic acid with water[1]

(1) Place 0.5 g of impure benzoic acid in a 25 mL beaker, add 10 mL of distilled water and several boiling stones. Heat the mixture to boiling on a heating mantle or hot plate and stir it with a glass rod to promote the dissolution. Add a small amount of hot water if necessary, keeping the mixture at the boiling, until all of the solute dissolves[2].

(2) If the solution is highly colored, just proceed to Step 3 described above.

(3) Remove the suspended solids by hot filtration.

(4) Cool the filtrate down to room temperature to crystallize the benzoic acid. Then separate the crystals from the mother liquor and wash them with a small amount of fresh and cold water for one or two times in terms of the rules mentioned above.

(5) Transfer and spread the crystals onto a watch glass. Then allow them to be dried under an infrared lamp[3].

(6) After completely drying, determine the melting point of the purified benzoic acid and compare it with that of impure product. Weigh the product and calculate the percentage yield.

2. Recrystallization of acetylaniline with water[4]

(1) Place 0.5 g of impure acetylaniline in a 25 mL beaker, and add an appropriate amount of distilled water[5] and several boiling stones. Heat the mixture at the boiling to dissolve all the solute[6].

(2) The following steps are the same to those of recrystallization of benzoic acid.

3. Recrystallization of acetylsalicylic acid with the mixture of ethanol and water

(1) Add 0.5 g of impure acetylsalicylic acid, 5～6 mL of 95% ethanol and several boiling stones into a 25 mL round-bottom flask fitted with a reflux condenser. Introduce the cooling water into the condenser and heat to boiling. Swirl the flask carefully to promote the dissolution.

(2) After completely dissolving, filter the solution to another flask by hot filtration[7], then fit the condenser and continue to heat at the boiling. Cautiously add, dropwise, the distilled water from the top of condenser until a slight turbidity is produced. Continue to heat to boiling, and the solution may be clear again. Gradually add the distilled water until the turbidity does not disappear even at the boiling.

(3) Allow the solution to cool down to room temperature and crystallize spontaneously.

(4) The following steps are the same to those of recrystallization of benzoic acid.

Ⅴ Notes and Instructions

【Notes】

[1] The solubility (in grams of solute per 100 mL of solvent) of benzoic acid in water: 0.21 g (17.5 ℃), 0.35 g (25 ℃), 2.2 g (75 ℃), 2.7 g (80 ℃), 5.9 g (100 ℃).

[2] Add 1～2 mL of hot water every time until there is no change for the quantity of the insoluble solids.

[3] The setting temperature of the infrared lamp should not exceed 100 ℃.

[4] The solubility of acetylaniline in water: 0.46 g(20 ℃), 0.56 g(25 ℃), 0.84 g(50 ℃), 3.45 g(80 ℃), 5.5 g(100 ℃).

[5] Distilled water (15～20 mL) is needed for 0.5 g of impure acetylaniline.

[6] "Oil" may be produced in the process of dissolving acetylaniline. It is not the impurity, but the eutectic formed by acetylaniline and water. It can be dissolved by adding more water and heating the mixture to boiling.

[7] Due to the high flammability of ethanol, the open flame or hot plate should be prevented.

【Requirements for preview】

(1) To know the basic principle and general process of recrystallization.

(2) To be familiar with the apparatus and operation requirements for hot filtration and vacuum filtration.

【Experimental precautions】

(1) If the organic solvent is used, dissolution of the solute must be carried out in an apparatus for reflux. Open flame and hot plate should be prevented when adding the solvent or removing suspended solids by hot filtration.

(2) Never add the activated charcoal into a boiling solution when decolorizing the solution.

(3) Hot filtration should be carried out quickly to avoid crystallization on the filter paper or in the funnel caused by the decreasing of temperature.

(4) When stopping the vacuum filtration, first disconnect the filter flask from the aspirator, or open the stopcock on the trap, then turn off the aspirator to avoid the suck-back of the water.

Ⅵ Post-lab Questions

(1) How many steps does the process of recrystallization involve? What is the purpose for each step?

(2) Why should the amount of the solvent consumed in recrystallization be neither too much nor too little? How to determine the consumption of the solvent correctly?

(3) What operations easily lead to a fire when using organic solvent in recrystallization? How about the correct operation?

(4) How to choose an appropriate solvent for recrystallization?

Ⅶ Verbs

recrystallization 重结晶；
hot filtration 热过滤；
solubility 溶解度；
vacuum filtration 减压过滤；
saturated solution 饱和溶液；
Büchner funnel 布氏漏斗；

filter flask 抽滤瓶；　　seed crystal 晶种；
solute 溶质；　　water aspirator 水泵

实验五　偶氮苯和苏丹(Ⅲ)的薄层色谱分离鉴定

一、实验目的

(1) 了解薄层色谱的基本原理和操作步骤。
(2) 掌握薄层色谱的鉴定方法和用途。

二、实验原理

薄层色谱又称薄层层析(thin-layer chromatography)，常用 TLC 表示。薄层色谱的特点是所需样品少(几微克到几十微克)，分离时间短(几分钟到几十分钟)，效率高，它是一种微量、快速和简便的分离分析方法，实验室最为常用。可用于精制样品、化合物鉴定、跟踪反应进程和柱色谱的先导，即为柱色谱摸索最佳条件等。

薄层色谱是将吸附剂均匀地涂在玻璃板或某些高分子薄膜上作为固定相，经干燥活化后点上待分离的样品，用适当极性的溶剂作为展开剂(即流动相)。当展开剂在吸附剂上展开时，由于样品中各组分对吸附剂吸附能力不同，发生连续的吸附和解吸过程，吸附能力弱的组分随流动相较快地向前移动，吸附能力强的组分则移动较慢。利用各组分在展开剂中溶解能力和被吸附剂吸附能力的不同，最终将各组分彼此分开。如果各组分本身有颜色，则薄层板干燥后会出现一系列高低不同的斑点；如果本身无色，则可用各种显色剂或在特殊光源下使之显色，以确定斑点位置。在薄板上混合物的每个组分上升的高度与展开剂上升的前沿之比称为该化合物的 R_f 值，又称比移值，见图 2-8。对于同一化合物，当实验条件相同时，其 R_f 值应是一样的。因此，可用 R_f 值来初步鉴定物质[1]。

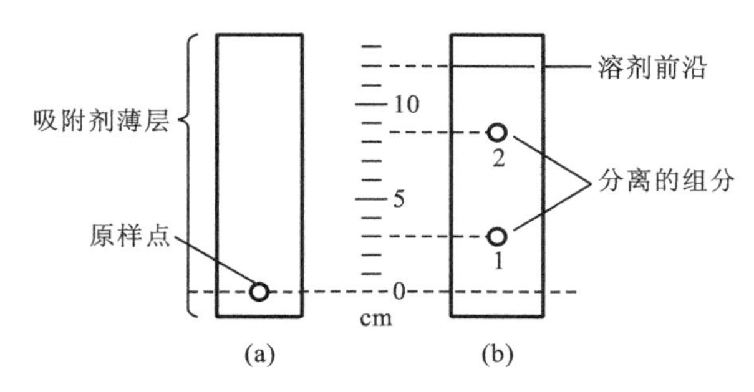

图 2-8　R_f 值的计算示意图

$$R_f=\frac{\text{溶质的最高浓度中心至原点中心的距离}}{\text{溶剂前沿至原点中心的距离}}$$

$$R_{f1}=\frac{3\ \text{cm}}{12\ \text{cm}}=0.25,\quad R_{f2}=\frac{8.4\ \text{cm}}{12\ \text{cm}}=0.70$$

1. 吸附剂

TLC 最常用的吸附剂是硅胶粉及氧化铝粉。硅胶是无定形多孔物质，略具酸性，适用于中性或酸性物质的分离，薄层色谱用的硅胶可分为以下几种。

(1) 硅胶 H:不含黏合剂和其他添加剂。

(2) 硅胶 G:含煅烧石膏($CaSO_4 \cdot 2H_2O$,做黏合剂)。

(3) 硅胶 HF254:含荧光物质,可在 254 nm 波长紫外光下观察荧光。

(4) 硅胶 GF254:含煅烧石膏及荧光物质。

与硅胶相似,氧化铝也因含黏合剂或荧光剂而分为氧化铝 G、氧化铝 GF254 及氧化铝 HF254。氧化铝的极性比硅胶的大,比较适合于分离极性小的化合物。

通常将薄层板按加黏合剂和不加黏合剂分为两种,加黏合剂的薄层称为硬板,不加黏合剂的薄层称为软板。常用的黏合剂除煅烧石膏外,还有淀粉、羧甲基纤维素钠(carboxymethyl cellulose,CMC)。化合物的吸附能力与它们的极性成正比,极性大则与吸附剂的作用强,随展开剂移动慢,R_f值小;反之,极性小则 R_f值大。因此,利用硅胶或氧化铝薄层色谱可把不同极性的化合物分开,甚至结构相近的顺、反异构体也可分开。各类有机化合物与上述两类吸附剂的亲和力大小次序大致为:羧酸>醇>伯胺>酯、醛、酮>芳香族硝基化合物>卤代烃>醚>烯烃>烷烃。

供薄层层析用的吸附剂粒度较小,通常为 200 目。当颗粒太大时,表面积小,吸附量少,样品随展开剂移动速度快,斑点扩散较大,分离效果不好;当颗粒太小时,样品随展开剂移动速度慢,斑点不集中,效果也不好。标签上有专门说明,不可和柱层析吸附剂混用。

2. 展开剂

展开剂的选择主要是根据样品的极性、溶解度和吸附剂的活性等因素。溶剂的极性越大,则对化合物解吸的能力越强,R_f值也越大。如果样品各组分 R_f值都较小,则可加入适量极性较大的溶剂混合使用。常用展开剂极性大小次序为:己烷和石油醚<环己烷<四氯化碳<三氯乙烯<二硫化碳<甲苯<苯<二氯甲烷<氯仿<乙醚<乙酸乙酯<丙酮<丙醇<乙醇<甲醇<水<吡啶<乙酸。

3. 薄层板的制备

薄层板通常为玻璃板,其制备得好坏直接会影响到分离效果,吸附剂应尽可能涂得牢固、均匀,厚度为 0.25~1 mm。实验室最常用的为湿板铺层的方法。

制湿板前首先要将吸附剂调成可流动的糊状浆料。

平铺法:用购置或自制的薄层涂布器进行制板,涂层既方便又均匀,是科学研究中大量制作时常用的方法。

倾注法:将调好的浆料倒在玻璃板上,用手摇晃振荡,使其表面均匀平整并铺满整板,然后放在水平的平板上晾干。这种制板方法厚度不易控制,适合于少量制板时使用。

晾干后的薄层板需要活化,硅胶板活化一般在 105~110 ℃烘 30 min。氧化铝活化在 200~220 ℃烘 4 h。活化后的薄层板放在干燥器内备用,以防吸湿失活,影响分离效果。

4. 展开槽

展开槽亦称层析缸,规格形式不一,有立式、卧式、斜靠式、下行式、上行式等。展开剂倒入层析缸中后,应待容器内溶剂蒸气达到饱和后,再将点好样的薄层板放入槽或缸中进行展开,展开时应加盖密闭(见图 2-9)。

5. 显色

样品展开后,如果本身带有颜色,可直接看到斑点的位置。但是,大多数有机物是无色的,因此就存在显色的问题。常用的显色方法如下。

图 2-9　TLC 展开方法

(1) 显色剂法。常用的显色剂有碘和三氯化铁水溶液等。许多有机化合物能与碘生成棕色或黄色的配合物。利用这一性质,在密闭容器中(一般用展开缸即可)放几粒碘,将展开并干燥的薄层板放入其中,稍稍加热,让碘升华,当样品与碘蒸气反应后,薄层板上的样品点处即可显示出黄色或棕色斑点,取出薄层板用铅笔将点圈好即可。除饱和烃和卤代烃外,均可采用此方法。三氯化铁溶液可用于带有酚羟基化合物的显色。

(2) 紫外光显色法。用硅胶 GF254 制成的薄板层,由于加入了荧光剂,在 254 nm 波长的紫外灯下,可观察到暗色斑点,此斑点就是样品点。

三、仪器与试剂

【仪器】滴管,载玻片,毛细管,层析缸,紫外灯,电吹风。

【试剂】乙酸乙酯-石油醚混合液(1∶2.5),偶氮苯的苯溶液,苏丹(Ⅲ)的苯溶液,偶氮苯和苏丹(Ⅲ)的混合液,二苯甲酮的苯溶液,乙酰苯胺的苯溶液,二苯甲酮和乙酰苯胺的混合液,1%的羧甲基纤维素钠(CMC)水溶液,硅胶 G,硅胶 GF254。

四、实验步骤

1. 制板

取医学载玻片 6 块,洗净,晾干。

在 50 mL 干燥小烧杯中,放入 1%羧甲基纤维素钠水溶液9 mL,逐渐加入 3 g 硅胶 G,调成均匀的糊状。用滴管吸取此糊状物,涂于上述洁净的载玻片上。用食指和拇指拿住载玻片,前后左右振荡摆动,使流动的硅胶 G 均匀地铺在载玻片上。一共制作 3 块硅胶 G 板。将涂好的硅胶 G 板水平置于实验台上,在室温下放置半小时或在平板电炉上垫石棉网约 40 ℃烘干固化。放至烘箱中,缓慢升温至 110 ℃,活化半小时后取出,稍冷后置于干燥器中备用。薄板制成后其涂层应无厚薄不匀、起泡、起层和脱落等现象,否则不能使用。

硅胶 G 板制作熟练后,在另一个 50 mL 干燥小烧杯中,放置 1%羧甲基纤维素钠水溶液 9 mL,逐渐加入 3 g 硅胶 GF254,调成均匀糊状,制成 3 块硅胶 GF254 板。其制作、活化等均同硅胶 G 板。

2. 点样

分别取少量 0.5%～1%偶氮苯和苏丹(Ⅲ)的苯溶液以及这两种化合物的二元混合液为试样。在离硅胶 G 板一端 1 cm 处,用铅笔在玻璃板两侧轻轻做记号。取管口平整的玻璃毛细管,插入样品溶液中,于铅笔记号中间水平处轻轻点样,样点直径约 3 mm。点样时,使毛细管下端液体刚好接触薄层即可,切勿点样过重造成凹陷而使薄层破坏。每块硅胶 G 板点两个样(一个标样和一个未知样),点样时点与点相距 1 cm 左右,一边点已知样,另一边点混合未知样(预留一块板备用)。

将二苯甲酮和乙酰苯胺的苯溶液以及这两种化合物的二元混合物[2],按上述步骤,在硅胶GF254 薄层板上点样。

3. 展开及定位

在层析缸中加入少量乙酸乙酯-石油醚混合液(1∶2.5)为展开剂,或用合适的烧杯或广口瓶代替层析缸,将薄层板点样一端朝下小心放入,注意展开剂液面高度不得超过点样线。盖上盖子,观察展开剂前沿上升到离板的上端约 1 cm 处时取出,不可让展开剂跑到薄板顶端。尽快用铅笔在展开剂上升的前沿画上记号,晾干。分别计算纯样和未知样中参照组分的 R_f 值,确定未知样组成。硅胶 GF254 板用紫外灯(254 nm 波长)进行观察,并用铅笔确定好斑点中心,计算 R_f 值。

五、注解和实验指导

【注解】

[1] 影响 R_f 值的因素很多,如样品的结构、吸附剂和展开剂的性质、薄层板的质量以及温度等。当这些实验条件都固定时,化合物的 R_f 值是一个特性常数。但由于实验条件很难达到完全相同,因此在鉴定一种具体化合物时,经常采用与已知标准样品对照的方法。

[2] 乙酰苯胺的极性大,受极性硅胶的吸附作用强,不易被极性较小的溶剂洗脱。

【预习要求】

(1) 了解色谱分析方法有哪些,各有什么用途和使用范围。

(2) 备好铅笔和直尺。想一想实验中将要记录哪些数据,预先设计一个表格,用来记录实验中的原始数据。

【操作注意事项】

(1) 点样用毛细管必须专用,不得弄混而引起样品污染。

(2) 薄层层析适合微量甚至痕量分析,上样过多会造成拖尾和追尾而无法达到分离的目的。

(3) 点样后吹干苯溶剂再展开,可以有效避免跑散、跑偏等现象。

(4) 层析缸内沿侧壁放置一张滤纸,盖上盖子,等滤纸被展开剂蒸气饱和后再放置载样玻片开始展开,对于得到效果较好的色谱图很有帮助。

(5) 从展开剂取出后立即在展开剂前沿画记号,如不注意,等展开剂挥发后,就无法确定展开剂上升的高度了。

六、思考题

(1) 展开剂的高度若超过了点样线,对薄层色谱有何影响?

(2) 在混合物薄层色谱中,如何判定各组分在薄层上的位置?如果斑点出现拖尾现象,这可能是什么原因所引起的?

(3) 如果色斑在层析薄板上无法随着溶剂快速移动,可以采取哪些措施?

Experiment 5　Separation of Azobenzene and Sudan(Ⅲ) by Thin-layer Chromatography

Ⅰ Objectives

(1) To understand the working principle of thin-layer chromatography(TLC)method.
(2) To separate and identify two components of a dye mixture by TLC.

Ⅱ Principle

Thin-layer chromatography(TLC) is a sensitive, fast, simple and inexpensive analytical technique that can be used repeatedly in carrying out organic experiments. It is a micro technique, as little as 10^{-9} g of material can be detected. TLC is a fast and effective method in the following applications: ① to determine the number of components in a mixture; ② to determine the identity of two substances; ③ to monitor the progress of a reaction; ④ to determine the effectiveness of a purification; ⑤ to determine the appropriate conditions for a column chromatographic separation; ⑥ to monitor column chromatography.

1. Working principle of TLC

TLC involves spotting the sample to be analyzed near one end of a sheet of glass or plastic that is coated with a thin layer of an adsorbent. The sheet, which can be the size of a microscope slide, is placed on end in a covered jar containing a shallow layer of solvent (see Figure 2-8). As the solvent rises by capillary action up through the adsorbent, the components of the mixture are partitioned between the mobile liquid phase and the stationary adsorbent phase. This is therefore a solid-liquid system, the stationary phase (the adsorbent) is solid and the mobile phase (the solvent, also called eluent) is liquid. The more strongly a given component of the mixture is adsorbed onto the stationary phase, the less time it will spend in the mobile phase and the more slowly it will migrate up on the TLC plate.

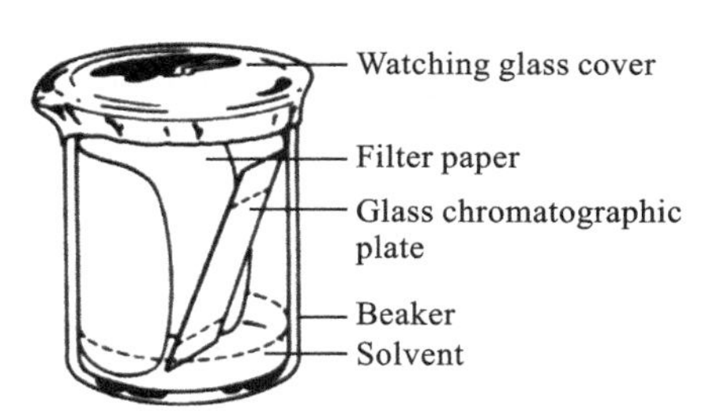

Figure 2-8　A Simple Method of Developing Thin-layer Chromatographic Plate

The migration of a certain substance can be characterized by its R_f value. This number is an index of how far a certain spot moved on the plate(see Figure 2-9). Using the R_f values, one is able to identify the components of the mixture with the individual compounds. The calculation of R_f value is as followes[1]:

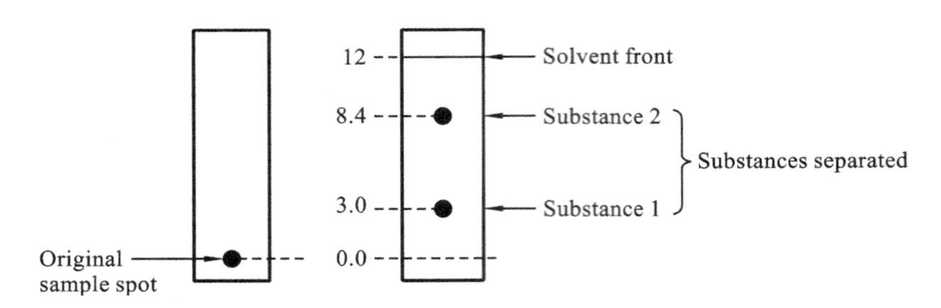

Figure 2-9　Calculation of R_f value on a TLC plate

$$R_f=\frac{\text{The distance from the origin to the center of a spot}}{\text{The distance from the origin to solvent front}}$$

$$R_{f1}=\frac{3.0\ \text{cm}}{12\ \text{cm}}=0.25,\quad R_{f2}=\frac{8.4\ \text{cm}}{12\ \text{cm}}=0.70$$

If the relative migration distance of one spot from the mixture is the same as that of a known compound for control when they are developing on the same plate, that is, $R_{f,\text{sample}}=R_{f,\text{control}}$, then this individual compound can be identified as one component of the mixture.

If the substance separated by TLC is colored, then it is possible to detect the components visually. Plates that have been impregnated with a fluorescent indicator will show dark spots for the compounds under an ultraviolet light.

2. Adsorbents and solvents

The two most common coatings for thin-layer chromatography plates are alumina, Al_2O_3, and silica gel, SiO_2, which are also widely used in column chromatography for the purification of macroscopic quantities of material. Of the two, alumina, when anhydrous, is the more active, i. e., it will adsorb substances more strongly. It is thus the adsorbent of choice when the separation involves relatively less polar substrates such as hydrocarbons, alkyl halides, ethers, aldehydes and ketones. To separate the more polar substrates such as alcohols, carboxylic acids and amines, the less active adsorbent, silica gel, is preferentially used.

In an extreme situation, very polar substances on alumina do not migrate very far from the starting point (give low R_f values) or nonpolar compounds travel with the solvent front (give high R_f values) if chromatographed on silica gel. These extremes of behavior are markedly affected, however, by the solvents used to carry out the chromatography. A polar solvent will carry along with polar substrates, and nonpolar solvents will do the same with nonpolar compounds—another example of the generalization "like dissolves like".

Ⅲ Apparatus, Reagents and Materials

【Apparatus and materials】Glass plates(6 pieces), blower, jar(or beaker), capillary tube.

【Reagents】Silica gel (G), fluorescent silica gel (GF254), 1% sodium carboxymethylcellulose (CMC) in distilled water, solvent mixture of ethyl acetate and petroleum ether in 1∶2.5;

Sample group(1): 1% azobenzene/sudan(Ⅲ)/the mixture of azobenzene and sudan(Ⅲ) in benzene;

Sample group (2): 1% acetanilide/benzophenone/the mixture of acetanilide and benzophenone in benzene.

Ⅳ Procedures

1. Preparing the TLC plates by coating

Place 3 g of silica gel G and 9 mL of 1% CMC solution in a 50 mL beaker, stir continuously to produce an even slurry. Swirl the flask to mix the slurry and draw a portion into a medicine dropper. And apply emulsion onto 3 pieces of glass plates one by one until the entire upper face is covered; shaking or tilting the slide to the left to cause a flow, and then to the right; tilt again to the front and to the rear until the emulsion cover the slide evenly. The adsorbent should be about 0.25 mm thick. Then place the finely covered glass plates on the table for a few minutes waiting for the flow dry. Then transfer the plates into oven for gradually drying for about 30 min under 40～110 ℃. Take them out of the hot oven and wait for cooling down before use.

Making other 3 pieces of TLC plates by coating silica gel GF254 following the same steps.

2. Spotting to the TLC plates

With a pencil, lightly draw a line about 1 cm from the edge. Insert the capillary into the solution that is to be analyzed and apply a tiny spot of the solution to the plate until it spreads to a spot of 1 mm diameter. Spot two samples on each plate (one is the known compound for control, the other is the mixture). For each sample use a separate capillary tube. Two equally spaced spots can get on the solid surface; the distance between each spot will be about 1 cm. Dry the spots and label each plate.

Spot the sample group (1) on silica gel(G)-coating plates; spot the sample group(2) on silica gel(GF254)-coating plates.

3. Developing the TLC plates

Pour about 10 mL of solvent mixture (ethyl acetate : petroleum ether=1 : 2.5) into a 100 mL beaker and place spotted plate diagonally for an ascending chromatography (see Figure 2-8). Make certain that the spots applied to the plate are above the surface of the eluting solvent. Cover the beaker with a watch glass to avoid the evaporation of the solvent mixture. When the solvent front nears the edge of the plate, about 1 cm from the edge, remove the plate from the beaker. You must not allow the solvent front to advance up to or beyond the edge of the plate. Mark immediately the position of the solvent front with a pencil and allow the solvent to dry.

4. Detecting the spots on the plates and calculate the R_f values

Mark the center of the spots and record the migration distance of the spots. Calculate the R_f values of each spot and identify the components in the mixture [2]. Examine the silica gel GF254 plates under UV-light to see the components as dark spots against a bright green-blue background. Outline the spots with a pencil. Record the distances and calculate the R_f values.

Ⅴ Notes and Instructions

【Notes】

[1] The R_f values are usually influenced by multiple experiment factors, such as the structure of sample, the properties of the adsorbents and solvents, the thickness and drying degree of the coated plates and temperature, etc.

[2] Acetanilide is much more polar than benzophenone, therefore the former is adsorbed by silica gel strongly and not easily moved along with solvent.

【Requirements for preview】

(1) To learn about the common chromatography methods in research work and their application field.

(2) Prepare a pencil and ruler. Think about what lab data must be recorded, and design a table for recording these data.

【Experimental precautions】

(1) When you make spots on the plate, you should keep the spots small and do not make a cave on the plate.

(2) Don't forget labeling the plates and not mix the plates coating by silica gel G with those by silica gel GF254.

(3) Before developing TLC plates, place a piece of filter paper around the inwall of the chamber and cover the chamber, wait for the solvent mixture wetting the whole paper and the chamber inside is filled with the solvent vapors, which will be helpful to maintain a stable solvent atmosphere in developing.

Ⅵ Post-lab Questions

(1) Why must the spot applied to a TLC plate be above the level of the developing solvent?

(2) What will be the result of applying too much sample on a TLC plate?

(3) What can you do when the spots cannot move along with the solvent on the plate?

Ⅶ Verbs

thin-layer chromatography 薄层层析(色谱);
adsorbent 吸附剂;
alumina 氧化铝;
silica gel G 硅胶 G;
fluorescent silica gel GF254
荧光硅胶 GF254;
chamber 展开槽;
capillary tube 毛细管;
blower 吹风机;
sodium carboxymethylcellulose(CMC)
羧甲基纤维素钠;
ethyl acetate 乙酸乙酯;
petroleum ether 石油醚;
azobenzene 偶氮苯;
sudan 苏丹;
acetanilide 乙酰苯胺;
benzophenone 苯甲酮

实验六　柱　色　谱

一、实验目的

(1) 了解柱色谱的原理及应用。

(2) 掌握柱色谱的操作步骤。

二、实验原理

柱色谱的基本原理与薄层色谱一样，利用混合物中各组分在固定相和流动相中分配系数的不同，从而达到分离的目的。

将已溶解的样品从柱顶加入，在柱的顶端被吸附剂(固定相)吸附，然后从柱顶加入有机溶剂做洗脱剂(流动相)，由于吸附剂对各组分的吸附能力不同，各组分以不同的速率下移。一般来说，吸附能力较强的组分在流动相中的质量分数比吸附能力较弱的组分要小，并以较慢的速率向下移动，各组分随洗脱剂以不同的时间从色谱柱下端流出。用容器分别收集洗脱液，然后除去溶剂，即可得到纯净的物质。若样品为有色物质，则可在色谱柱上形成肉眼可见的分段的色带。

柱色谱是分离、提纯和鉴定有机化合物的重要方法，有着极其广泛的用途。在有机合成实验中，要得到高纯度的有机化合物，产物往往需要进行柱色谱分离才能达到要求。在实际操作过程中，柱色谱常与薄层色谱配合使用。利用薄层色谱跟踪反应进程，优化柱色谱分离条件，并监测柱色谱分离情况。

1. 吸附剂

常用的吸附剂有硅胶、氧化铝、氧化镁、碳酸钙和活性炭等。实验室一般用氧化铝或硅胶，其中氧化铝的极性更大，是一种高活性和强吸附性的极性物质。市售的氧化铝通常分为中性、酸性和碱性三种，酸性氧化铝适用于分离酸性有机物质；碱性氧化铝适用于分离碱性有机物质，如生物碱和烃类化合物；中性氧化铝应用最为广泛，适用于分离中性物质，如醛、酮、酯、醇等。市售的硅胶略带酸性。

由于样品是吸附在吸附剂表面的，因此颗粒大小均匀、比表面积大的吸附剂分离效率最佳。比表面积越大，组分在固定相和流动相之间达到平衡就越快，色带就越窄。通常使用的吸附剂颗粒大小以100～150目为宜。

吸附剂的活性还取决于吸附剂的含水量。含水量越高，活性越低，吸附剂的吸附能力就越弱；反之，则吸附能力强。吸附剂的含水量与活性等级的关系如表2-4所示。

表2-4　吸附剂的含水量和活性等级的关系

活性等级	Ⅰ	Ⅱ	Ⅲ	Ⅳ	Ⅴ
氧化铝含水量/(%)	0	3	6	10	15
硅胶含水量/(%)	0	5	15	25	38

常用的是Ⅱ级和Ⅲ级吸附剂。Ⅰ级吸附性太强，而且易吸水；Ⅳ级吸水性弱；Ⅴ级吸附性太弱。

2. 洗脱剂

洗脱剂的选择一般是通过薄层色谱实验来进行的。具体方法如下:先用少量溶剂溶解样品,在已制备好的薄层板上点样。能将样品中各组分完全分开的溶剂,即可作为柱色谱的洗脱剂。若单一溶剂达不到所要求的分离效果,可考虑选用混合溶剂做洗脱剂。

选择洗脱剂的另一个原则是:洗脱剂的极性不能大于样品中任一组分的极性。否则会由于洗脱剂在固定相上被吸附,迫使样品一直保留在流动相中。在这种情况下,各组分在柱中移动得非常快,很难建立起分离所要达到的吸附-解吸平衡,从而影响分离效果。

不同的洗脱剂使给定的样品沿着固定相的相对移动能力,称为洗脱能力。在硅胶和氧化铝柱上,洗脱能力按下列溶剂顺序提高:石油醚、己烷、环己烷、甲苯、二氯甲烷、氯仿、乙醚、乙酸乙酯、丙酮、正丙醇、乙醇、甲醇、水。

三、仪器与试剂

【仪器】色谱柱,固体加料漏斗,滴液漏斗,锥形瓶。

【试剂】硅胶G,中性氧化铝(100～200目);1%苏丹(Ⅲ)和对硝基苯胺的混合苯溶液,1%邻硝基苯胺和对硝基苯胺的混合苯溶液,1∶2的乙酸乙酯-石油醚混合液,95%乙醇,含有1 mg荧光黄与1 mg亚甲基蓝的10 mL 95%乙醇溶液,0.05%甲基橙和0.25%亚甲基蓝混合乙醇溶液。

【物理常数】

试　剂	相对分子质量	密度/(g·cm^{-3})	熔点/℃	沸点/℃	溶解性	
					H_2O	醇
苏丹(Ⅲ)	352.39	—	199(分解)	—	不溶	溶
对硝基苯胺	138.13	1.42	146～148	332	不溶	溶
邻硝基苯胺	138.13	1.44	73～76	284.5	溶于热水	溶
亚甲基蓝	319.85	—	190～191	—	溶	微溶
荧光黄	332.31	—	314～316	—	不溶	溶于热乙醇
甲基橙	327.94	1.28	300	—	微溶	不溶

四、实验步骤

方案一　苏丹(Ⅲ)和对硝基苯胺的分离

本实验中吸附剂为硅胶,洗脱剂为1∶2的乙酸乙酯-石油醚混合液,苏丹(Ⅲ)的极性较小,首先被洗脱出来,而极性较大的对硝基苯胺则需要用更大极性的乙醇(或氯仿)才能顺利洗脱。

1. 色谱柱的准备和安装

由于水会改变洗脱液的极性,所用色谱柱必须是干燥的。使用前首先用少量洗脱液对色谱柱检漏,必须保证活塞处没有渗漏且溶剂流通顺畅。将长度约为20 cm的小色谱柱垂直安装,下面用25 mL小锥形瓶作为洗脱液接收器。用镊子取少量脱脂棉(或玻璃毛)放在干净的色谱柱底部。在脱脂棉上盖上一层0.5 cm厚的石英砂或者大小合适的滤纸片(见图2-10)。

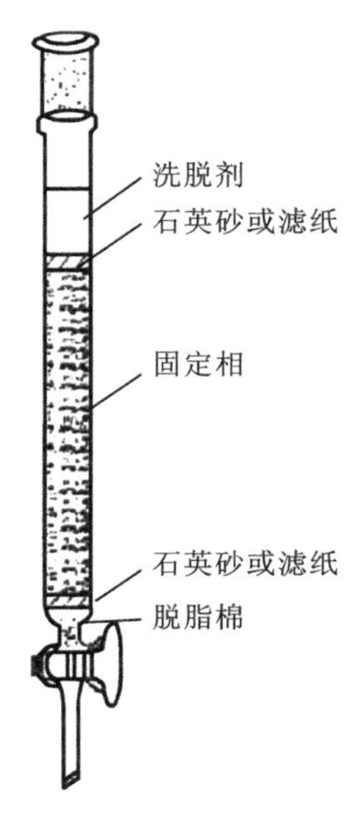

图 2-10　色谱柱

关闭活塞，向柱内倒入 1∶2 的乙酸乙酯-石油醚混合液至柱高的 2/3 处。称取 12～18 g 的硅胶，置于干燥的烧杯中，加入约 30 mL 洗脱液，用玻璃棒或刮抄搅拌硅胶至均匀分散。打开色谱柱活塞，控制流速为每秒 1 滴。柱顶加一个干燥的固体加料漏斗，边搅拌边将硅胶和洗脱液的混合物加入色谱柱，让湿润的硅胶在溶剂中自由沉降进行填装[1]。当硅胶装至柱高 3/4 处时，在上面加一层 0.5 cm 厚的石英砂或者加盖一张大小合适的滤纸片。操作时，注意保持流速，液面不能低于石英砂上层[2]。

2. 上样和洗脱

当溶液液面刚好流至石英砂面时，关闭活塞，用滴管沿柱壁加入 1 mL苏丹(Ⅲ)和对硝基苯胺的混合苯溶液。开启活塞，当液面即将流至石英砂面时再次关闭活塞，用少量洗脱液洗下壁管残留的有色物质，然后在色谱柱顶部装上装有 50 mL 1∶2 的乙酸乙酯-石油醚混合液的滴液漏斗，等所有有色残留液刚好全部渗入固定相时，打开滴液漏斗活塞开始洗脱[3]。

3. 样品收集和鉴定

观察洗脱过程中形成的色带。当红色色带快洗出时更换接收器，继续淋洗至滴出液为无色。更换接收器，改用乙醇(约 30 mL)为洗脱剂，当黄色液开始洗出时，用另一接收器收集，至黄色物质洗下为止。

分别将红色接收液和黄色接收液用旋转蒸发仪蒸出大部分溶剂，然后转移至蒸发皿，用红外灯烘干，固体干燥后测熔点。

方案二　邻硝基苯胺和对硝基苯胺的分离

同上装好色谱柱，当 1∶2 的乙酸乙酯-石油醚混合液恰好流至石英砂面时，用滴管沿柱壁加入 1 mL 邻硝基苯胺和对硝基苯胺的混合苯溶液。当溶液面降至硅胶上端表面时，用滴管加入洗脱液洗下黏附在管壁的混合物，并打开活塞使所有有色残留液刚好全部渗入吸附剂中，反复几次直至洗净。然后装上滴液漏斗，用 1∶2 的乙酸乙酯-石油醚混合液淋洗，控制速度，当黄色的邻硝基苯胺达到柱底时，更换接收器，接收该色带，然后收集淡黄色的对硝基苯胺色带。

分别将黄色接收液和淡黄色接收液用旋转蒸发仪蒸出大部分溶剂，冷却结晶，固体干燥后测熔点。

方案三　荧光黄(或甲基橙)和亚甲基蓝的分离

装柱前应先将砂芯色谱柱洗干净，进行干燥。关闭活塞。向柱中加入 10 mL 95% 乙醇，打开活塞，控制流速为每秒 1～2 滴。此时从柱上端通过长颈漏斗缓慢加入 5 g 中性氧化铝，用橡皮塞或手指轻轻敲打柱身下部，使填装紧密[4]。上面再加一层 0.5 cm 厚的石英砂。整个过程中一直保持乙醇流速不变，并注意保持液面始终高于吸附剂氧化铝的顶面。

当洗脱剂液面刚好流至石英砂面时，立即沿柱壁加入 1 mL 含有 1 mg 亚甲基蓝与 1 mg 荧光黄的 95% 乙醇溶液(或 0.5 mL 的 0.05%甲基橙和 0.25%亚甲基蓝混合乙醇溶液)。开至最大流速。当加入的溶液流至石英砂面时,立即用0.5 mL 95% 乙醇洗下管壁的有色物质,如此2~3次,直至洗净为止。

加入10 mL 95% 乙醇进行洗脱。亚甲基蓝首先向柱下移动,荧光黄(甲基橙)则留在柱上端,当第一个色带快流出来时,更换接收瓶,继续洗脱。当洗脱液快流完时,应补加适量的95%乙醇。当第一个色带快流完时,不要再补加95% 乙醇,等到乙醇流至吸附剂液面时,轻轻沿壁加入1 mL 水,然后加满水。取下此接收瓶进行蒸馏,回收乙醇。更换接收瓶接收第二个色带,直至无色为止。

五、注释与实验指导

【注解】

[1] 色谱柱填装紧密与否对分离效果很有影响,若松紧不均,特别是有断层或气泡时,会影响流速和色带的均匀性,色谱柱填装过紧,则使流速太慢。此种湿法装柱可以有效防止断层和气泡的产生。

[2] 为了保持柱子的均一性,使整个吸附剂浸泡在溶剂或溶液中是必要的。否则当柱中溶剂或溶液流干时,就会使柱身干裂,影响渗滤和显色的效果。

[3] 如不安装滴液漏斗,也可用每次倒入10 mL 洗脱剂的方法进行洗脱。

[4] 此种半湿法装柱尤其要注意不可将吸附剂填装过紧,并防止气泡和断层。

【预习要求】

(1) 了解色谱法的原理和应用。

(2) 弄清薄层色谱与柱色谱的区别与联系。

(3) 理解柱色谱中洗脱剂和吸附剂极性的选择与样品洗脱的关系。

【操作注意事项】

(1) 加入石英砂的目的是使加料时不会把吸附剂冲起,影响分离效果。若无石英砂,也可用大小合适的滤纸覆盖在吸附剂表面,注意放置时保持吸附剂表面平整。若吸附剂表面被扰动,可以用玻璃棒将上层硅胶搅动后重新沉降成水平平面。

(2) 柱色谱分离过程中应一直保持洗脱剂的流速不变,并注意保持液面始终高于吸附剂的顶面。

(3) 色带从色谱柱被洗脱出来之前,接收的洗脱液可以倒回滴液漏斗中重复使用。

六、思考题

(1) 为什么必须保证所装柱中没有气泡?

(2) 洗脱时为什么要先用非极性或弱极性的洗脱剂,然后用较强极性的洗脱剂?

(3) 邻硝基苯胺和对硝基苯胺的极性哪个更大?哪个先被洗脱出来?为什么?

Experiment 6 Column Chromatography

Ⅰ Objectives

(1) To understand the basic principle of column chromatography.

(2) To learn how to operate the column chromatography.

Ⅱ Principle

The basic principle of column chromatography (CC) is same to thin-layer chromatography(TLC). It can separate and purify the mixture on the basis of the difference in partition coefficient between the stationary phase and mobile phase for each component.

When a solution containing dissolved samples is put in the fixed column, the samples will be adsorbed on the adsorbent (stationary phase). Then the eluents (mobile phase) are added to elute the samples to form different bands on the adsorbent. Generally, the more strongly a given component of the mixture is adsorbed on the stationary phase, the less time it will spend in the mobile phase and the more slowly it will move downward in the column.

Column chromatography is an important method to separate, purify and identify the organic compounds. In many organic synthesis experiments, only column chromatography can give the products with high purity. And in these procedures, TLC is also needed to monitor the progress of a reaction, determine the appropriate conditions of CC and monitor the process of CC.

1. Adsorbents

It is very important to choose suitable adsorbents as stationary phase. Silica gel, alumina, magnesium oxide, calcium carbonate and activated carbon are commonly used as adsorbents in CC. In organic chemistry lab, silica gel and alumina are preferable to be used to purify the mixture of organic compounds, and the polarity of the former is lower than that of the latter. There are three kinds of commercial alumina, such as neutral, acidic and basic. Acidic alumina can be used to separate acidic organic compounds; basic alumina is suitable for basic organic compounds, such as alkaloids and hydrocarbons; neutral alumina is of wide-ranging applications, which can be used to separate aldehydes, ketones, esters and alcohols etc. The commercial silica gels are slightly acidic.

Since samples are adsorbed on the surface of adsorbents, its dimensions and specific surface area will affect the effectiveness of separation greatly. The larger the specific surface area is, the more quickly the balance between the stationary and mobile phase for each component can be achieved and the narrower the bands are. The particle size of the adsorbents commonly used is 100～150 meshes.

Meanwhile the activity of adsorbents depends on the water content. The higher the water content, the lower the activity will be. The relationship between activity grades and water contents is listed in Table 2-4.

Table2-4 Water Contents and Activity Grades of Adsorbents

Activity Grade	Ⅰ	Ⅱ	Ⅲ	Ⅳ	Ⅴ
Water in Alumina/(%)	0	3	6	10	15
Water in Silica Gel/(%)	0	5	15	25	38

In these different adsorbents, grade Ⅱ and Ⅲ are widely used. However grade Ⅰ is of high adsorb ability and hygroscopicity, and grade Ⅳ and Ⅴ contain too much water resulting in low adsorb ability.

2. Eluents

The strength of adsorption depends upon the compounds involved. Since the adsorbents are polar, compounds with high polarity will be adsorbed more tightly and non-polar compounds are firstly eluted. Eluents used as mobile phases in CC can be determined by TLC experiments. To do so, normally, a separation will begin by using non-polar or low-polarity solvent, allowing the compounds adsorbed to the stationary phase, then slowly switching the polarity of the solvent to desorb the compounds and allow them to travel with the mobile phase. The polarity of the solvents should be changed gradually. Sometimes mixture of solvents can also be considered as a choice.

The principle to choose eluents is the polarity of solvent. If the solvent is much more polar than the compounds, the three-way equilibrium will shift in favor of the solvent-adsorbent interactions. The compounds will remain in the mobile phase, and separation will not occur. Alternatively, if the compounds are much more polar than the solvent, the three-way equilibrium will shift in favor of the sample-adsorbent interactions. No compounds will be eluted since the solvent is unable to move compounds from the adsorbent sites.

Different solvents have different abilities to elute samples. The commonly used solvents are shown as follows in the increasing order of polarity: Petroleum ether, hexane, cyclohexane, toluene, dichloromethane, chloroform, ethyl ether, ethyl acetate, acetone, *n*-propanol, ethanol, methanol and water.

Ⅲ Apparatus and Reagents

【Apparatus】 Chromatographic column(dry), solid-feeding funnel(dry), dropping funnel (dry), Erlenmeyer flasks(dry).

【Reagents】 Silica gel, neutral alumina; 1% Sudan(Ⅲ) and 1% *p*-nitroaniline in benzene; 1% *o*-nitroaniline and 1% *p*-nitroaniline in benzene; ethyl acetate ∶ petroleum ether(1 ∶ 2); 95% ethanol; the solution of methylene blue and sodium fluorescein (1 mg of each dissolved in 10 mL of 95% ethanol), mixture solution of 0.05% methyl orange and 0.25% methylene blue in ethanol.

【Physical constants】

Reagent	M_w	ρ/ ($g \cdot cm^{-3}$)	m. p. /℃	b. p. /℃	Solubility	
					In H_2O	In EtOH
Sudan Ⅲ	352.39	—	199(dec.)	—	insoluble	soluble
p-Nitroaniline	138.13	1.42	146～148	332	insoluble	soluble
o-Nitroaniline	138.13	1.44	73～76	284.5	soluble(hot)	soluble
Methylene blue	319.85	—	190～191	—	soluble	slightly soluble
Sodium fluorescein	332.31	—	314～316	—	insoluble	soluble(hot)
Methyl orange	327.94	1.28	300	—	slightly soluble	insoluble

Ⅳ Procedures

Scheme 1 Separation of Sudan (Ⅲ) and *p*-Nitroaniline

The adsorbent used in this experiment is silica gel. The eluent is mixture of ethyl acetate and petroleum (1 : 2), and ethanol (or chloroform) is used to elute *p*-nitroaniline.

1. Preparing the column

(1) Turn off the stopcock of chromatographic column to be used. Add a small amount (2～3 cm high)of eluent solvent in the column to check if it's leaking at the stopcock. Report to your guider or grease the stopcock after dried. It is important to make sure the column would not be leaking during the eluting.

(2) Vertically fix a 20 cm dry column in iron support with a 25 mL Erlenmeyer flask as receiver. Put some degreasing cotton (or glass wool) in the clean bottom of the column with a tweezer, and then cover the cotton with 0.5 cm high silica sand (or use a piece of filter paper cut in proper size) (see Figure 2-10).

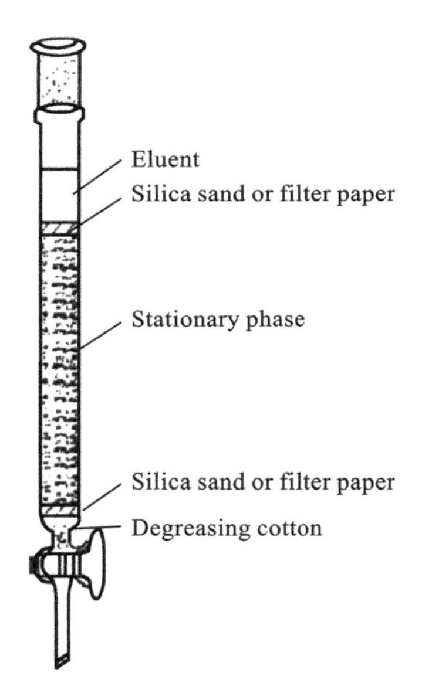

Figure 2-10 Chromatographic Column

(3) Turn off the stopcock, fill the column 2/3 full with the eluent (ethyl acetate : petroleum ether=1 : 2). Weigh 12～18 g of silica gel in a beaker, add 30 mL of eluent and stir the mixture to keep the silica gel well distributed in the solvent.

(4) Turn on the stopcock, and let the eluent flow down at the rate of 1 drop per second. Slowly add the slurry into the column with a dry solid-feeding funnel, let the wet silica gel drop to the bottom in the solvent until it fills about 3/4 full of the column[1]. Note that liquid level should be higher than the surface of silica get during the whole process. Finally turn off the stopcock and cover the silica gel with 0.5 cm high silica sand (or a piece of filter paper).

2. Loading the sample

(1) Maintain the flow rate of eluent. When the liquid is just above silica sand, turn off stopcock, carefully drop 1 mL of benzene solution of 1% Sudan (Ⅲ) and 1% *p*-nitroaniline along the inwall of the column.

(2) Now keep the liquid sample flow into the silica gel through the silica sand at a moderate rate. When all the liquid near to the surface of silica sand, stop the flowing and wash down the residue adhering to the inwall with a small amount of eluent and keep all the residue infiltrate the adsorbents. Place a dropping funnel at the top of column, and pour 50 mL of eluent into the funnel.

3. Eluting the sample

Turn on the stopcock, maintain a moderate flow rate to start eluting from the top[2]. Observe the separation of a red band and a yellow band. Start to receive the first colored band of eluent with a dry Erlenmeyer flask when the band approaches to the stopcock. And replace the receiver for the second. The blank band of eluent in the transition also should be collected in another receiver.

4. Identifying the sample

Evaporate the solvent in the collected liquids with rotary evaporator respectively, and then transfer the products to evaporation dish. Dry the solids under a infrared lamp and determine their melting points.

Scheme 2 Separation of *o*-Nitroaniline and *p*-Nitroaniline

Prepare chromatographic column according to the aforementioned method. When the liquid level drops to the surface of silica sand, transfer 1 mL benzene solution of 1% *o*-nitroaniline and 1% *p*-nitroaniline into the column along the inwall with a dropper. When the liquid gets close to the surface of silica sand, wash down the mixture adhering to the inwall with eluent, and then let the residual mixture infiltrate the adsorbents. Place the dropping funnel on top of the column, and then elute the column with ethyl acetate : petroleum ether(1 : 2). The yellow *o*-nitroaniline liquid will be collected firstly, and then the flavescent *p*-nitroaniline liquid.

Evaporate most of the solvents with rotary evaporator, cool down to crystallize, and determine the melting points of products after drying.

Scheme 3 Separation of Sodium Fluorescein (or Methyl Orange) and Methylene Blue

Wash and dry the chromatographic column before packing. Turn off the stopcock; pour 10 mL of 95% ethanol into the column, then turn on the stopcock, maintaining the flow rate at 1～2 drops per second. Add neutral alumina(5 g) to the column slowly through a long-stem funnel, and then tap the column softly with stick or rubber plug in glass rod to fill the alumina tightly. Finally cover the alumina with silica sand(0.5 cm in height). Keep liquid flowing constantly.

When the liquid is just above the silica sand, transfer 1 mL of 95% ethanol solution, which containing 1 mg methylene blue and 1 mg sodium fluorescein(or 0.5 mL 0.05% methyl orange and 0.25% methylene blue ethanol solution) into the column along the inwall with a dropper. When the liquid gets close to silica sand, wash down the mixture adhering to the inwall with 0.5 mL of 95% ethanol for 2～3 times.

Add 10 mL of 95% ethanol to elute the column. Methylene blue will flow out firstly, while sodium fluorescein(methyl orange) still stays immovable. When the blue band comes out, change a new receiver to collect it. If the eluent is used up, pour a little more 95%

ethanol into the column. When the blue band is finished, wait the ethanol level down to silica sand and change water as eluent. Add 1 mL of water along the inwall softly, then fill the column full, collect yellow sodium fluorescein solution. Remove most of the ethanol in methylene blue solution with rotary evaporator.

Ⅴ Notes and Instructions

【Notes】

[1] The adsorbent in chromatographic column should be filled in uniform tightness. If it isn't so, especially when with small fault and bulb, the colored band will be badly-distributed and the flow rate be affected greatly. But note that, if it is filled too much tightly, the flow rate will be very slow. This method for filling the column is good to avoid too much tightness and some faults.

[2] During the separation, the flow rate should be kept constantly, and the eluent level should always be higher than the surface level of adsorbents.

【Requirements for preview】

(1) To understand the fundamental principle of chromatography and its applications.

(2) To understand the differences and relationship between TLC and CC.

(3) To clear about how to choose the eluents and absorbents with different polarity.

【Experimental precautions】

(1) Silica sand is to help fix the surface level of adsorbents; it can be replaced by a filter paper with proper sized to keep the adsorbents stable at the surface.

(2) All adsorbents should be kept uniformly and immersed in solvents during the elution.

(3) The eluents received from the bottom of column can be repeatedly used before the colored band comes out of the column.

Ⅵ Post-lab Questions

(1) Why can't air bubbles be permitted into column?

(2) Why should lower-polar solvents be prior to higher-polar solvents when eluting?

(3) Which one is more polar, *o*-nitroaniline or *p*-nitroaniline? Which one is eluted prior to the other and why?

Ⅶ Verbs

column chromatography 柱色谱；　　stationary phase 固定相；

eluent 洗脱剂；　　mobile phase 流动相

实验七　纸　色　谱

一、实验目的

(1) 学习纸色谱的基本原理。

(2) 掌握纸色谱分离氨基酸的操作技术。

二、实验原理

纸色谱类似于薄层色谱,但是其原理属于分配色谱。纸色谱以滤纸作为载体,滤纸纤维吸附着一定量的水作为固定相,移动相采用水与一种或多种有机溶剂混溶的混合溶剂,称为展开剂。

将点样的滤纸条下端浸入展开剂中,由于滤纸纤维的毛细管作用,展开剂将向上移动。展开剂到达样品点后,溶解样品共同向上移动,而固定相则要溶解的样品停留,于是样品便在移动相和固定相间进行分配。样品的不同组分中,亲水性强的组分容易留在固定相中,于是随展开剂上升的速度较慢;亲脂性强的组分则上升速度较快,从而实现不同组分的分离。衡量物质向上移动的物理量是比移值 R_f,其计算过程详见实验五。

在一定的条件下,某种物质的 R_f 值是常数。R_f 值的大小与物质的结构、性质、溶剂系统、层析滤纸的类型和层析温度等因素有关,因而一般采用在相同实验条件下与对照物质对比的方法来确定其异同。本实验利用纸色谱法分离氨基酸。

三、仪器与试剂

【仪器】色谱缸(干燥大试管,配软木塞),喷雾器,电吹风,刻度尺(20 cm),滤纸条(1.5 cm×16 cm),铅笔,木框(16 cm×40 cm),大头针,图钉,毛细管,棉线。

【试剂】样品:氨基酸(蛋氨酸、谷氨酸、丙氨酸)混合液[1];

展开剂:被水饱和的苯酚溶液[2];

显色剂:0.2%茚三酮乙醇溶液。

四、实验步骤

1. 制备滤纸条

将滤纸条平放于洁净的纸上,在距离滤纸下端 1 cm 处用铅笔轻画一横线,并在横线的中央画一个直径为 2 mm 的小圆圈。

2. 点样

用毛细管吸取待分离的氨基酸混合液,在滤纸条下端的小圆圈内沾一下,让混合液流入纸条(切勿滴入过多,以免扩散至小圆圈外),任其在空气中晾干。如图 2-11(a)所示,在样品点下端用大头针别住,使滤纸条垂直;在滤纸条上端穿一个小孔,小孔上牵一根棉线,棉线两端左右分开。

3. 展开

如图 2-11(b)所示,将滤纸条垂直悬于装有被水饱和的苯酚溶液的试管内[3],勿使滤纸条接触试管壁。调节线的高度使滤纸下端浸入苯酚溶液 0.5 cm,塞好软木塞,注意勿使样品点浸入溶液中,否则展开剂会溶解样品或使样品斑点严重扩散以致无法正常展开进行色谱分离。此时即看到溶剂向上移动,溶剂升到离滤纸条上端约 5 cm 时,小心将纸条取出,用手指捏住上端,以铅笔标出溶剂上升的前沿。

氨基酸样品展开过程大约 1 h。

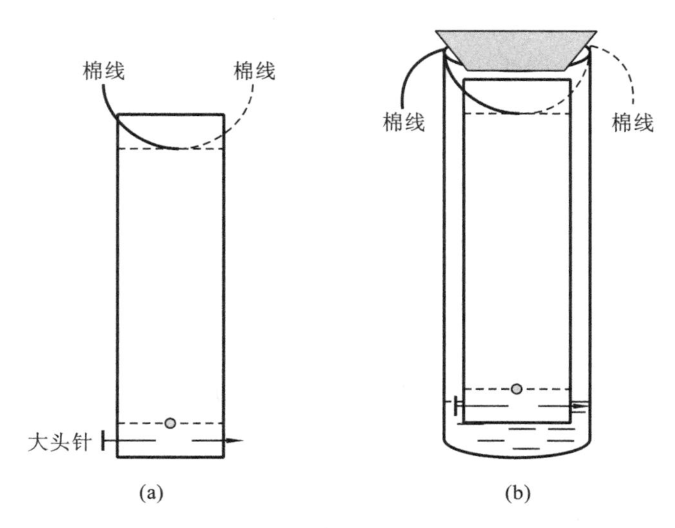

图 2-11　纸色谱实验装置示意图

4. 显色

将滤纸条两端用图钉钉在小木框上，用电吹风吹干后，用喷雾器均匀地喷薄薄一层 0.2% 茚三酮乙醇溶液，然后再将滤纸条以热风吹干，即可看到纸上显出三个色点，每个色点代表一种氨基酸。

5. 计算 R_f 值

用刻度尺分别测量点样点（即原点）至各色斑点中心的距离，再测量点样点至溶剂前沿的距离，计算各色点的 R_f 值，根据三种氨基酸的 R_f 参考值（蛋氨酸 0.84，谷氨酸 0.32，丙氨酸 0.62）确定此三个色点分别为何种氨基酸。

五、注解和实验指导

【注解】

[1] 氨基酸混合液：将蛋氨酸、谷氨酸各 3 mg 与丙氨酸 2 mg 溶于 0.5 mL 蒸馏水中即成，制备后须放在冰箱内保存，并尽可能当天配制。

[2] 展开剂（即被水饱和的苯酚溶液）：取新鲜蒸出的苯酚约 50 g 及蒸馏水 30 mL 放在分液漏斗内，用力振摇后，放置 7～10 h，使分成两层，取出下层使用。

[3] 在干燥的大试管中，加入被水饱和的苯酚溶液 10 mL，注意勿使溶液沾到试管壁上，塞好塞子，将试管垂直固定于铁架台上，饱和 20 min。

【预习要求】

(1) 了解各种氨基酸的结构。

(2) 学习纸色谱原理及其与薄层色谱、柱色谱之间的关系。

(3) 准备铅笔、直尺、计算器。

【操作注意事项】

(1) 点样量要适当，太少则效果不明显，太多则分离不好。

(2) 画线时只能使用铅笔，不能使用其他笔。其他笔的颜色为有机染料，在有机溶剂中染料溶解，颜色会产生干扰。

(3) 无论是画线还是点样，不能用手接触滤纸条前沿线以下的任何部位，因为手指上有相当量的氨基酸，在本实验中会产生干扰。

六、思考题

(1) 纸色谱的展开剂中，为什么要含有一定比例的水?
(2) 悬挂滤纸条时为什么不能接触试管壁?
(3) 展开结束后，进行显色时，为什么要先烘干再喷显色剂?
(4) 为什么各氨基酸 R_f 值不同?

Experiment 7 Paper Chromatography

Ⅰ Objective

(1) To understand the basic principle of paper chromatography.
(2) To learn the separation method of amino acids by paper chromatography.

Ⅱ Principle

Paper chromatography is often considered to be related to thin-layer chromatography (TLC). The experimental techniques are somewhat like those of TLC, but the principles are more closely related to those of extraction. Paper chromatography is actually a liquid-liquid partitioning technique rather than a solid-liquid technique.

For paper chromatography, the support is filter paper. The stationary phase is water absorbed on the paper, and the moving phase (or the development solvent) is organic solvents that contain water as a component.

In paper chromatography, a small spot of sample is placed near the bottom of a piece of filter paper. Then the paper is placed in a developing chamber. The developing solvent ascends on the paper by capillary action and the components of the spotted mixture move upward at differing rates. Because the water phase is stationary, the components in a mixture that are more highly water-soluble are held back and move more slowly. On the contrary, the more lipophilic components are less held and ascend more quickly. As the solvent ascends on the paper, the compounds are repeatedly partitioned between the stationary water phase and the moving solvent and travel upward. The ration of the distance the compound travels to the distance the solvent front travels is called the retention factor value(R_f). The R_f value of a compound can be calculated as indicated in Experiment 5.

When the conditions of measurement are completely specified, the R_f value is constant for any given compound. Its value is related to the structure and property of the compound, developing solvent and type of filter paper, temperature, etc. Hence, the R_f value can be used to identify unknown compounds by comparing them with standard substances whose identity and R_f value are known under the same experiment condition. Here paper chromatography is used to separate and identify amino acids.

Ⅲ Apparatus and Reagents

【Apparatus】Developing chamber (a big test tube with a rubber stopper), spray bottle, hair dryer, ruler(20 cm), filter paper strip(1.5 cm×16 cm), pencil, wood frame(16 cm×40 cm), pin, thumbtack, capillary tube, cotton thread.

【Reagents】Sample: the aqueous solution of methionine, glutamic acid and alanine mixture[1];

Developing solvent: the solution of phenol saturated by water[2];

Visualization reagent: 0.2% of ninhydrin solution dissolved in ethanol.

Ⅳ Procedures

1. Marking a filter paper strip

Take a piece of filter paper strip. Use a pencil to draw a baseline about 1 cm from the bottom of the filter paper strip, whose dimension is 5 cm×20 cm. Mark one point in the middle as starting on the baseline and draw a circle with 2 mm diameter around the point.

2. Spotting the filter paper strip

Dip the capillary tube into solution of amino acid to be examined, the solution rises up to the capillary tube by capillary action. Then make the loaded capillary tube touch the filter paper strip lightly at the point marked on the baseline. The capillary tube should be touched to the paper strip very briefly and then removed. The solution is transferred to the paper strip as a small spot. It is best for the spot not to spread out of the circle with 2 mm diameter. Blow gently on the paper to dry the spot. Then the filter paper strip is pierced through by a pin on the bottom so that it can hang in the chamber straightly. Prick a hole in the paper strip near the top with a pin and then pull a cotton thread through the hole(see Figure 2-11(a)).

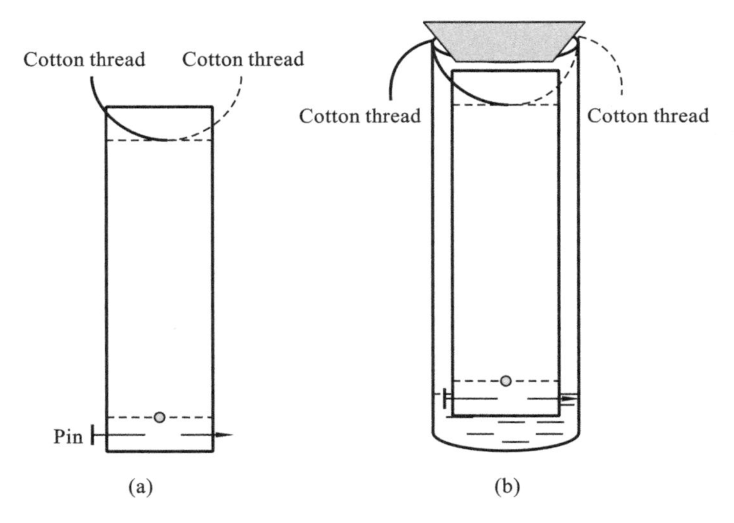

Figure 2-11　Assembling a Developing Chamber of Paper Chromatography with a Filter Paper Strip Hanging

3. Developing the filter paper strip

As shown in Figure 2-11(b), the paper strip is put into a big test tube with 10 mL of the solution of phenol saturated with water[3], by which it hangs freely and does not touch the wall of the chamber. Then adjust the cotton thread to allow the bottom of the paper strip to dip into the solvent about 0.5 cm depth. It is important that the level of the solvent in the bottom of the chamber must not be above the spot that was applied to the filter paper, or the spotted material will dissolve in the pool of solvent or diffuse seriously instead of undergoing chromatography. Close the chamber with a rubber stopper and then the solvent front climbs up. When the solvent has advanced to within 5 cm of the end of the filter paper, the paper strip may be taken out carefully of the chamber, and the position of the solvent front should be marked immediately with a pencil. The developing process carries out nearly 1 h.

4. Visualization

Fix the ends of the filter paper strip on a wood frame with two thumbtacks and blow it to dry with a hair dryer. Spray 0.2% of the ninhydrin solution on the paper strip. The paper should be uniformly moist but not too wet. Blow it to dry again, the spots begin to appear. When the spots are sufficiently intense, remove the dryer and stop heating. We can see three purple spots on the paper strip. Each spot represents one amino acid.

5. Calculating the R_f value

Outline the spots and mark a point at the center of each spot with a pencil. Measure the distance that the compound has traveled from the point at which it was originally spotted. Then measure the distance that the solvent front has traveled from the same original spot. Calculate the R_f value of various amino acids. Determine the assignment of amino acids according to the reference R_f value.

The reference R_f value of methionine, glutamic acid and alanine is 0.84, 0.32 and 0.62, respectively.

The experiment requires 4 h.

Ⅴ Notes and Instructions

【Notes】

[1] Amino acid aqueous solution is prepared by dissolving the mixture of 3 mg of methionine, 3 mg of glutamic acid, and 2 mg of alanine in 0.5 mL of distilled water. The sample should be saved in the refrigerator. It is best for the sample to be used once prepared.

[2] Developing solvent, the solution of phenol saturated with water, is prepared by shaking the solution of 50 g of fresh distilled phenol and 30 mL of distilled water in the separatory funnel and standing for 7～10 min to allow the phases to separate. Drain the lower layer through the stopcock to use.

[3] Place 10 mL of the solution of phenol saturated with water in a big dry test tube and close the chamber with a rubber stopper. Don't make the solution touch the wall of the test tube. Fix the tube vertically on the iron stand and keep the chamber saturated with the

solvent vapors for about 20 min.

【Requirements for Preview】

(1) Know about the structure of various amino acids.

(2) Learn about the principle of paper chromatography and the relation of itself, thin-layer chromatography and column chromatograph.

(3) Prepare a pencil, ruler and calculator.

【Experimental precautions】

(1) It is important to spot the suitable quantity of sample into the filter paper strip. It may be no spot to appear with less sample and difficult to separate with too much sample.

(2) Only pencil can be used to draw for paper chromatography, or organic dye of other pens may dissolve in the pool of the solvent to interfere with chromatographic separation.

(3) Don't touch any part of the filter paper strip under the solvent front with your hands when you draw a line or spot the sample. Because the hands have a large number of amino acids, it will interfere with chromatographic separation.

Ⅵ Post-lab Questions

(1) Why should the development solvent contain some water in paper chromatography?

(2) Why can't the hung filter paper touch the wall of the test tube in the chromatography chamber?

(3) Developed filter paper should be dried before visualization and then spray visualization reagents on the paper strip. Why?

(4) Why are the R_f values of various amino acids different?

Ⅶ Verbs

paper chromatography 纸色谱；
development chamber 层析缸；
methionine 蛋氨酸；
glutamic acid 谷氨酸；
alanine 丙氨酸；
phenol 苯酚；
ninhydrin(水合)茚三酮

实验八　萃　　取

一、实验目的

(1) 掌握液-液萃取的原理。

(2) 熟悉分液漏斗的使用。

二、实验原理

萃取是提取、分离或纯化有机化合物的常用操作之一。按萃取两相的不同，萃取可分为液-液萃取、液-固萃取。液-液萃取是利用同一物质在两种互不相溶(或微溶)的溶剂中具有不同溶解度的性质，将其从一种溶剂转移到另一种溶剂中，从而达到分离或提纯目的的一种方法。

能斯特分配定律是液-液萃取方法的理论依据。在一定温度下,同一种物质在两种互不相溶的溶剂(A 和 B)中遵循如下分配原理:

$$\frac{C_A(\text{物质在 A 溶剂中的浓度})}{C_B(\text{物质在 B 溶剂中的浓度})}=K$$

式中,K 为分配系数。在萃取时,提高分配系数可以提高萃取的效率。

分液漏斗是一种玻璃实验器皿,常见的为球形和梨形,可用于互不相溶的液体的分离。其主要结构包括斗体、盖在斗体上口的塞子和在斗体下口具两通结构的活塞(见图 2-12)。将分液漏斗洗净、检漏。确认不漏水后,关好活塞,将分液漏斗置于固定在铁架台上的铁圈中,把待萃取混合液(体积为 V)和萃取剂(体积约为 $V/3$)倒入分液漏斗,盖好上口塞。用右手握住分液漏斗上口,并以右手食指摁住上口塞;左手握住分液漏斗下端的活塞部位,小心振荡,使萃取剂和待萃取混合液充分接触(见图 2-13)。振荡过程中,要不时将漏斗尾部向上倾斜并打开活塞,以排出因振荡而产生的气体。振荡、放气操作重复数次后,将分液漏斗再置于铁圈中,静置分层。当两相分清后,先打开分液漏斗上口塞,然后打开活塞,使下层液体经活塞孔从漏斗下口慢慢放出,上层液体自漏斗上口倒出。这样,萃取剂便带着被萃取物质从原混合物中分离出来。一般像这样萃取三次即可。

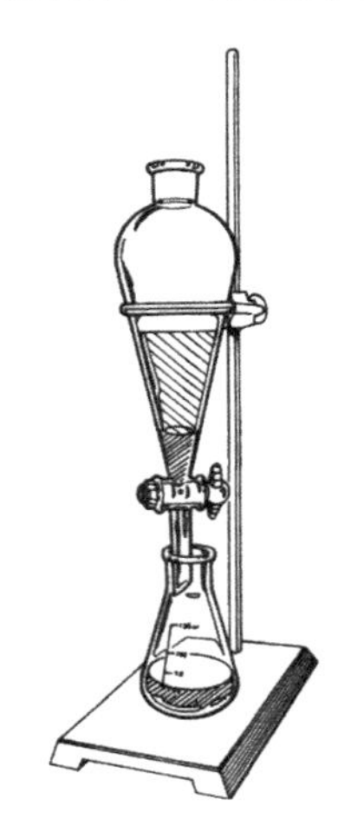

图 2-12 分液漏斗的构造

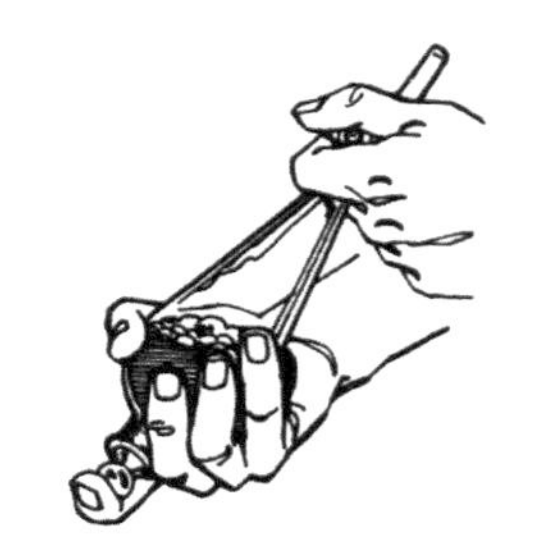

图 2-13 分液漏斗的振荡操作

三、仪器与试剂

【仪器】60 mL 分液漏斗,锥形瓶,移液管,洗耳球,碱式滴定管。

【试剂】2% 乙酸溶液,乙酸乙酯,氢氧化钠,酚酞。

【物理常数】

名　称	相对分子质量	密度/($g \cdot cm^{-3}$)	溶解度/[$g \cdot (100\ mL)^{-1}$]		
			水	乙醇	乙醚
乙酸乙酯	88.11	0.897	8.3	混溶	混溶
乙酸	60.05	1.049	混溶	混溶	混溶

四、实验步骤

1. 一次萃取

用移液管准确移取 2%乙酸溶液 5.00 mL 于 60 mL 分液漏斗中[1]。另取 14 mL 乙酸乙酯加入分液漏斗，振荡萃取，朝无人处放气[2]，于铁圈中静置分层。先打开分液漏斗上口塞，然后通过下方活塞将下层水溶液分离至 250 mL 锥形瓶中，加 5 mL 水，用 NaOH 标准溶液滴定，记下消耗标准溶液的体积(mL)。分液漏斗中上层的乙酸乙酯从上口倒出，回收。

2. 分次萃取

在 60 mL 分液漏斗中准确加入 2% 乙酸溶液 5.00 mL，再加入 7 mL 乙酸乙酯进行萃取，将下层水溶液分离至另一洁净的分液漏斗中，再加入 7 mL 乙酸乙酯萃取一次。将下层溶液分离至 250 mL 锥形瓶中，加 5 mL 水，加入 1～2 滴酚酞做指示剂，用 NaOH 标准溶液滴定，记下消耗标准溶液的体积(mL)，分液漏斗中上层的乙酸乙酯从上口倒出，回收。计算萃取率，并比较用乙酸乙酯 14 mL 一次萃取和每次 7 mL 分两次萃取的萃取率。

$$萃取率=\left[1-\frac{c_{\mathrm{NaOH}}(\mathrm{mol\cdot L^{-1}})\times V(\mathrm{mL})}{1000\times n_{乙酸}}\right]\times 100\%$$

五、注解和实验指导

【注解】

[1] 由于乙酸溶液后期要进行滴定，必须尽量准确量取，所以要用移液管进行操作。用左手拿洗耳球，右手握移液管并用食指控制液面至刻线。

[2] 萃取过程中，振荡后放气时须朝无人处，以避免腐蚀性的气体或液体冲出造成伤害。

【预习要求】

(1) 学习相似相溶原理。

(2) 学习移液管和碱式滴定管的正确使用方法。

(3) 了解分液漏斗使用前如何检漏。

【操作注意事项】

(1) 分液漏斗使用之前必须用水检渗漏，如果有渗漏，应该在打开活塞后擦干活塞及活塞处内壁附着的水珠，然后在活塞上孔的两侧涂上适量的凡士林，小心插入活塞后顺着一个方向旋转几圈直到顺滑透亮。涂好后再一次检漏，直到确保不渗漏。

(2) 在打开下方的活塞释放下层液体时，需先打开分液漏斗上方的塞子，或者将塞子的孔槽与瓶口的孔槽对齐。

(3) 上层液体必须从分液漏斗的上方倒出，以免受到污染。

六、思考题

(1) 为什么乙酸乙酯能把乙酸从水溶液中萃取出来？

(2) 萃取时，如果有机物的密度未知，如何准确判断有机相在上层还是下层？

Experiment 8 Extraction

Ⅰ Objectives

(1) To understand the principle of liquid-liquid extraction.

(2) To be familiar with the usage of separatory funnel.

Ⅱ Principle

Extraction is an important part of the purification and separation methods. Transferring a solute from one solvent to another is called liquid-liquid extraction. The solute is extracted from one solvent into the other by the distribution. Any liquid substance with similar polarity (either polar or non polar) dissolves into each other, so called "like dissolves like". When a solution is shaken with a second solvent with which it is immiscible, the solute distributes itself between the two liquid phases. When the two phases have separated again into two distinct solvent layers, equilibrium will have been achieved so that the ratio of the concentrations of the solute in each layer defines a constant, which is called the distribution coefficient K. At equilibrium,

$$K=\frac{C_1}{C_2}$$

where C_1 and C_2 are the concentrations of the solute A in solvent 1 and in solvent 2, respectively. This relation is independent of the total concentration and the actual amount of the two solvents mixed.

A separating funnel takes the shape of a cone with a hemispherical end. It has a stopper at the top and stopcock(tap)at the bottom(see Figure 2-12). Check the separatory funnel and make sure that it is not leaking by adding some water in the funnel. Pour the solution to be separated in the funnel from the top. The funnel is then closed and shaken gently by inverting the funnel multiple times(see Figure 2-13). During the shaking, the stopcock should be turned

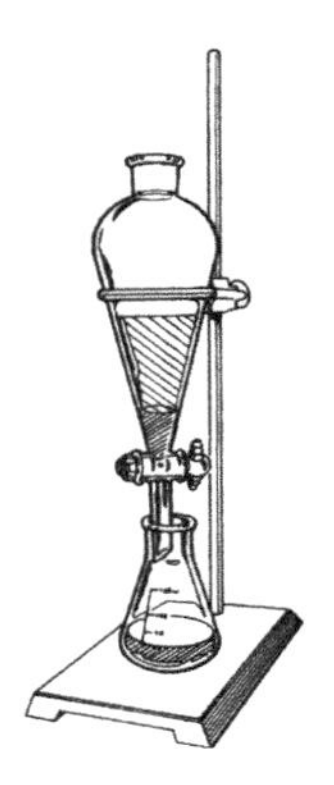

Figure 2-12 Separatory Funnel for Separation

Figure 2-13 Correct Positions for Holding a Separatory Funnel when Shaking

on carefully with the funnel slightly inverted to release excess vapor pressure, which is caused by quick evaporation of the organic solvent under shaking. The separatory funnel is set aside to allow for the complete separation of two phases. The lower layer will be removed through the stopcock. The upper layer must be swilled from the top of the funnel for avoiding of contamination. Generally, three times of extraction will get an effective result.

Ⅲ Apparatus and Reagents

【Apparatus】Separatory funnel(60 mL), Erlenmeyer flask, transfer pipet, alkali burette, rubber pipet bulb.

【Reagents】2% acetic acid solution, ethyl acetate, sodium hydroxide, phenolphthalein.

【Physical constants】

Reagent	M_w	ρ /(g · cm^{-3})	Solubility/[g · (100 mL)$^{-1}$]		
			Water	Ethanol	Diethyl ether
Ethyl acetate	88.11	0.897	8.3	Misible	Misible
Acetic acid	60.05	1.049	Misible	Misible	Misible

Ⅳ Procedures

1. Check for leaks

Firstly turn off the stopcock of separatory funnel. Add some distilled water in the funnel before use to make sure there is no leaking at the stopcock or the cap. If leaks occurred, take the stopcock out of the funnel, clean and dry it, then grease it with some vaseline and set up again (Too much grease will clog the hole in the stopcock and also contaminate the extract!), rotate it clockwisely for several times until it looks clear and feels smooth. Check for another time with water until no leaks occur any more.

2. One-time extraction

1) Extraction

(1) Check to make sure that the stopcock is closed and transfer 5.00 mL of acetic acid accurately with a transfer pipet into the separatory funnel well-prepared, then add 14 mL of ethyl acetate[1]. (The funnel should not be filled with more than three-fourths of its height.)

(2) Hold the separatory funnel correctly(see Figure 2-13)and shake it thoroughly with pressure released often[2].

(3) Fixed the separatory funnel on the iron clamp(ring)(see Figure 2-12)and allow the layers to separate completely and then draw off the lower layer into a beaker (What does this layer contain?) and pour the upper layer out through the neck to a 250 mL Erlenmeyer flask.

2) Titration(Check content of acetic acid)

(1) Add 5 mL of water and 1～2 drops of phenolphthalein solution to an Erlenmeyer flask.

(2) Place standard NaOH solution into an alkali burette, make sure that the tip of the burette and the rubber tube are full of NaOH solution. Then begin to titrate. When the color

of the liquid becomes light pink and lasts for 30, the end point arrives.

(3) Record the volume of NaOH solution which has been used.

(4) Calculate extraction rate by the equation as follows:

$$\text{extraction rate} = \left[1 - \frac{c_{NaOH}(\text{mol} \cdot \text{L}^{-1}) \times V(\text{mL})}{1000 \times n_{acid}}\right] \times 100\%$$

where n_{acid} is the mole number of 5.00 mL acetic acid.

3. Fractional extraction

1) Extraction

Place 5.00 mL of acetic acid to a separatory funnel, add 7 mL of ethyl acetate into the funnel. Shake the funnel thoroughly and turn on the stopcock with the stem tilt up and point away from people to release pressure often during the shaking. Allow the layers to separate completely, and draw off the lower layer into a beaker. Add an additional 7 mL of ethyl acetate to the separatory funnel, shake the mixture and draw off the lower layer as before, pour the up layer into a 250 mL Erlenmeyer flask(What does the flask contain?).

2) Titration

Add 5 mL of water and 1～2 drops of phenolphthalein solution in the Erlenmeyer flask and titrate the solution with standard NaOH solution as described above. Record the volume of NaOH solution which has been used and calculate extraction rate.

Compare the extraction rate between one-time extraction and fractional extraction. Which operation is more effective?

Ⅴ Notes and Instructions

【Notes】

[1] Because the acetic acid solution will be titrated at last, we should use transfer pipet to measure the volume for better accuracy. The transfer pipet is held by right hand and the pipette bulb is pressed for several times by left hand, then the liquid will be sucked into the pipet. Loose your forefinger and the liquid level will descend. When the liquid level reaches the marked line, press your forefinger tightly and outpour the liquid into another flask.

[2] Turn on stopcock cautiously (with the funnel stem pointed away from nearby people) to release pressure.

【Requirements for preview】

(1) Review the "like dissolves like" principle.

(2) Review the operation of titration.

(3) Get to know how to use a separatory funnel and how to check for leaking.

【Experimental precautions】

(1) The separatory funnel must be checked for leaking before use.

(2) When the lower layer is drawn off from below, the separatory funnel should be well held and the stopper at the neck should be taken off. Otherwise, the liquid cannot flow smoothly through the stopcock.

(3) The upper layer must be poured from the top of the funnel for avoiding of contamination.

Ⅵ Post-lab Questions

(1) Why can ethyl acetate extract acetic acid from the water solution?

(2) If the density of the colorless solute to be extracted is unknown, how can you recognize which layer is organic layer after extraction?

Ⅶ Verbs

separatory funnel 分液漏斗；　　titration 滴定

extraction 萃取；

实验九　从茶叶中提取咖啡因——萃取和升华

一、实验目的

(1) 学习天然物质中有效成分的提取方法和应用。

(2) 掌握索式提取器的原理及其应用。

(3) 掌握液-固萃取、蒸馏、升华的联合操作。

二、实验原理

茶叶中含有多种黄嘌呤衍生物的生物碱，咖啡因(caffeine，又名咖啡碱)的含量为1%～5%，并含有少量茶碱和可可豆碱，以及11%～12%的单宁酸(又称鞣酸)，还有约0.6%的色素和蛋白质等。咖啡因的化学名为1,3,7-三甲基-2,6-二氧嘌呤。其结构式如下：

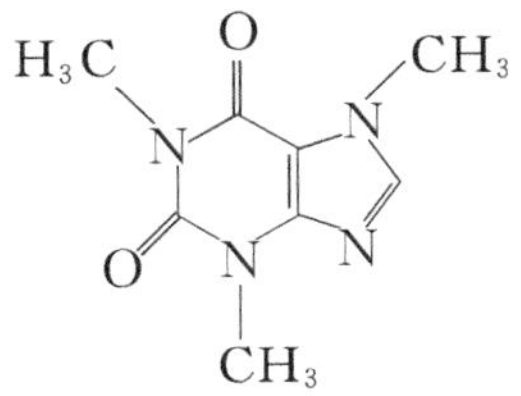

H_3C　O　CH_3
N　N
O　N　N
CH_3

纯咖啡因为白色针状晶体，无臭，味苦，置于空气中有风化性。咖啡因具有刺激心脏、兴奋大脑神经和利尿等作用，因此可作为药物使用。过度饮用咖啡因会增加抗药性并产生轻度上瘾。

咖啡因易溶于水、乙醇、氯仿、丙酮，微溶于石油醚，难溶于苯和乙醚，它是弱碱性物质，其水溶液对石蕊试纸呈中性反应。咖啡因在100 ℃时失去结晶水并开始升华，120 ℃时升华显著，178 ℃时升华很快。无水咖啡因的熔点为238 ℃。

咖啡因可由人工合成法或提取法获得。本实验采用索氏提取法从茶叶中提取咖啡因。利用咖啡因易溶于乙醇、易升华等特点，以95%乙醇做溶剂，通过索氏提取器进行连续抽提，然后浓缩、焙炒，再通过升华得到纯的咖啡因[1]。

升华法是精制某些固体化合物的方法之一。其基本原理是具有较高蒸气压的固体物质，在其熔点温度以下加热[2]，不经过液态直接变成蒸气，蒸气遇冷后又直接凝华变成固体。能用升华方法提纯的物质必须满足两个条件：

(1) 被提纯的固体要有较高的蒸气压(在室温下高于20 mmHg)；

(2) 杂质的蒸气压应与被提纯固体化合物的蒸气压之间有显著的差异。

具有较高蒸气压的物质才适宜用升华法来提纯,升华法得到的产品纯度较高,但有时损失较大。升华可以在常压下进行,也可以减压升华。

三、仪器、材料与试剂

【仪器】方法一:索氏提取器,平底烧瓶,球形冷凝管,蒸馏头,直形冷凝管,尾接管,锥形瓶,200 ℃温度计,蒸发皿,大三角漏斗。

方法二:500 mL 烧杯,500 mL 分液漏斗,50 mL 圆底烧瓶,常压蒸馏装置。

【材料】棉花,圆形滤纸,大头针。

【试剂】方法一:茶叶 2 g,95%乙醇,氧化钙粉末 4 g。

方法二:茶叶 25 g,碳酸钠 20 g,二氯甲烷 50 mL,无水硫酸镁,丙酮,石油醚(60~90 ℃)。

【物理常数】

化合物	熔点/℃	升华点/℃	溶解度/[g·(100 mL)$^{-1}$]			
			水	乙醇	丙酮	氯仿
咖啡因	235~238	178	2.2(25 ℃) 18.2(80 ℃) 66.7(100 ℃)	1.5(25 ℃) 4.5(60 ℃)	2(25 ℃)	18.2(25 ℃)

四、实验步骤

方法一　索氏提取法

1. 提取

按索氏提取器装置图安装提取装置(见图 2-14)。在提取装置的烧瓶中,放入两粒沸石,再装上索氏提取器。用滤纸做成与索氏提取器大小相适应的滤纸筒,在筒内放入约 2 g 碎茶叶,置入索式提取器中,滤纸筒上端要高于虹吸管顶端但不能堵住蒸气侧管口[3]。通过玻璃漏斗缓慢加入 95%乙醇,以刚好产生一次虹吸为准,然后多加约 10 mL,最后装上球形冷凝管回流。经加热回流约 1 h,产生 3~4 次虹吸,至索氏提取器中提取液颜色较淡时为止。最后一次虹吸后,立即停止加热。

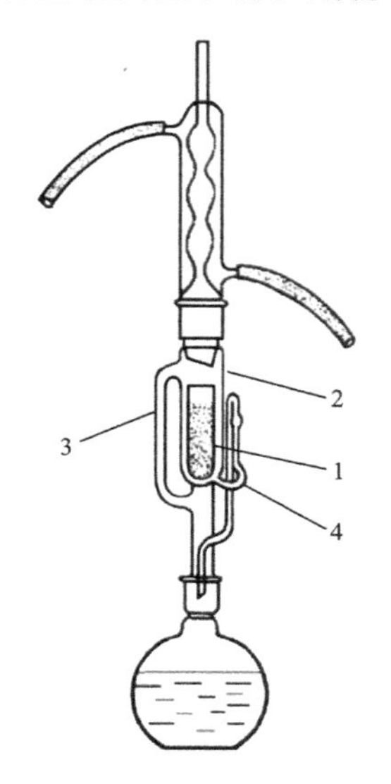

图 2-14　索氏提取器装置图

1—滤纸套;2—索氏提取器;3—蒸气侧管;4—虹吸管

2. 蒸馏

稍冷后取下回流冷凝管和索式提取器,改成普通蒸馏装置,用常压蒸馏法蒸出大部分溶剂。当蒸馏烧瓶中还剩下约 10 mL 液体时停止加热。不可过于浓缩,否则浓稠糊状物难以全部转出,造成损失。

3. 升华

将烧瓶中的浓缩液趁热倒入蒸发皿中,温火继续使溶剂挥发浓缩至约 5 mL,呈较稀薄糊状,加入 2 勺研细的生石灰粉末,拌匀。将蒸发皿放在加热器上缓慢加热成干燥粉末状。此时应十分注意加热强度,并充分翻搅、研磨。既要确保炒干,又要避免过热升华损失。若蒸发皿内物不粘玻璃棒，成为松散的粉末，表示溶剂已基本除去。稍冷后小心用滤纸擦去粘在蒸发皿边壁上的粉末，以免污染产物。然后进行常压升华操作。

在蒸发皿上面盖一张刺有许多小孔的滤纸。然后把一个直径略小于蒸发皿的玻璃漏斗倒置在上面，用一团棉花堵塞玻璃漏斗的颈部（见图 2-15），在石棉网上加热蒸发皿，逐渐升高温度至 200 ℃左右，继续升华 20 min。咖啡因蒸气通过滤纸孔，遇到冷的漏斗内壁，再凝结为晶体，附在漏斗的内壁和滤纸上。用穿有小孔的滤纸隔在中间可防止升华后形成的晶体落回到下面的蒸发皿中，以及吸收一些挥发性有色物质。本实验操作的关键是控制加热温度。如温度太高，有色杂质挥发，会使产品变色，咖啡因也会烤焦；温度太低，样品会在蒸发皿内壁上结出，与残渣混在一起。待能观察到玻璃漏斗内有可见的白色晶体时，停止加热。稍冷后收集白色咖啡因的针状或网状晶体。

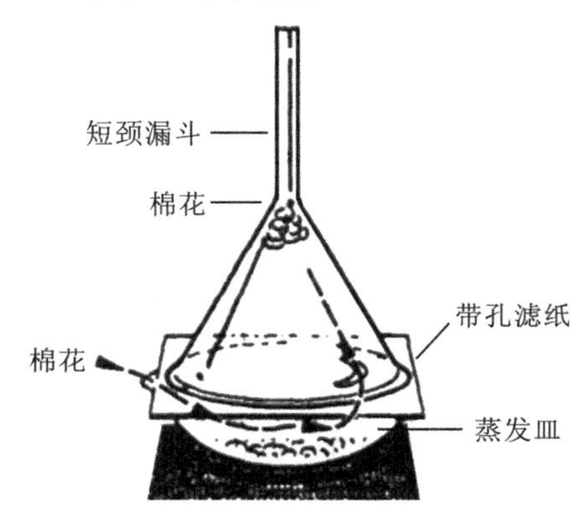

图 2-15　常压升华装置图

方法二　浸　取　法

在 500 mL 烧杯中，将 20 g 碳酸钠溶于 250 mL 蒸馏水，配成溶液。称取 25 g 茶叶，用纱布包好后放入烧杯中煮沸 0.5 h，注意勿使溶液起泡溢出。取出茶叶，压干，趁热抽滤。滤液冷至室温后，转入 500 mL 分液漏斗。加入 50 mL 二氯甲烷振摇 1 min，静置分层，此时在两相界面处产生乳化层。在小玻璃漏斗的颈口放置一小团棉花，棉花上放置约 1 cm 厚的无水硫酸镁，从分液漏斗直接将下层的有机相滤入干燥的锥形瓶，并用 2～3 mL 二氯甲烷淋洗干燥剂。水相再用 50 mL 二氯甲烷萃取一次，分层后通过重新加入的干燥剂。如过滤后的有机相混有少量的水，可重复上述操作一次，收集于锥形瓶中的有机相应是清亮透明的。

将干燥后的有机相分批转入 50 mL 圆底烧瓶，加入几粒沸石，在水浴上蒸馏回收二氯甲烷，并用水泵将溶剂抽干。含咖啡因的残渣用丙酮-石油醚重结晶。将蒸去二氯甲烷的残渣溶于最少量的丙酮，慢慢向其中加入石油醚（60～90 ℃），到溶液恰好混浊为止，冷却结晶，用玻璃钉漏斗抽滤收集产物。干燥后称重并计算收率。

五、注解和实验指导

【注解】

[1] 通过测定熔点及红外光谱、核磁共振谱等鉴定咖啡因，也可使之与水杨酸作用生成水杨酸盐（熔点为 137 ℃）以做确证。

[2] 物质的熔点温度可近似看作其三相点温度，高于三相点温度时，固态物质将会经过液态转化成气态，而不再是升华了。

[3] 受热的蒸气从侧管上升，遇冷后冷凝滴回提取腔内，当达到虹吸管的高度时，腔内液体会在重力作用下全部沿虹吸管虹吸回到下端的烧瓶内，这就是索式提取器的虹吸原理。

【预习要求】

(1) 了解提取的基本原理及其在医药学研究中的意义。

(2) 复习回流装置及普通蒸馏装置的搭建方法。

【操作注意事项】

(1) 使用索氏提取器时应十分注意保护侧面的虹吸管,勿使其碰破。

(2) 茶叶最好研碎,可以提高萃取效率,但要注意包好,防止碎屑堵塞虹吸管。

(3) 改为蒸馏装置后勿忘添加沸石。

(4) 蒸馏后若残留液过于黏稠,可将蒸出的乙醇倒回少量到烧瓶,尽量将黑色残留液全部转入蒸发皿中进行升华。

(5) 升华时温度控制尤其关键。可固定一支温度计测浴温来大致判断蒸发皿内的温度,可在 150 ℃下升华 15 min 后停止加热,冷却后收集一次产物,然后将皿内的残渣拌和均匀后再升至 200 ℃继续升华 15～20 min,使之升华完全。

六、思考题

(1) 索氏提取装置要垂直安装,为什么?

(2) 升华较重结晶在应用上有哪些优点和局限性?

Experiment 9　Isolation of Caffeine from Tea —Extraction and Sublimation

Ⅰ Objectives

(1) To demonstrate the isolation of a natural product.

(2) To be familiar with the techniques of extraction and simple distillation.

(3) To use sublimation as a purification technique.

Ⅱ Principle

Caffeine is one of the main substances of tea leaves. Besides being found in tea leaves, caffeine is present in coffee, kola nuts and cocoa beans. As much as 5% by weight of the leaf material in tea plants consists of caffeine.

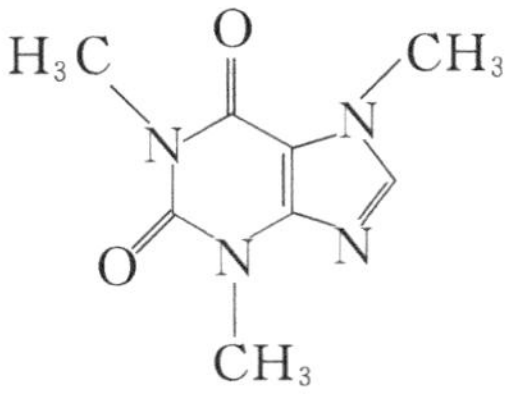

The caffeine structure is shown above. It is classed as an alkaloid, meaning that with the nitrogen present, the molecule has base characteristics(alkali-like). In addition, the molecule has the purine ring system, a framework which plays an important role in living systems.

1. Soxhlet extractor for continuous extraction

Many organic compounds are obtained from natural sources through extraction. In this experiment, caffeine is readily soluble in hot organic solvent—ethanol and is thus separated from the tea leaves. For separation of the components of a solid mixture by continuous solid-liquid extraction, a Soxhlet extraction apparatus(see Figure 2-14)is convenient to escape of tedious work of a very large number of extraction with smaller quantities of solvent. The solid is placed in a porous thimble in the chamber, as shown, and the extracting solvent is in the boiling flask below. The solvent is heated to reflux, and the distillate, as it drops from the condenser, collects in the chamber. By coming in contact with the solid in the thimble, the liquid effects the extraction. After the chamber fills to the level of the upper reach of the siphon arm, the solution empties, from this chamber into the boiling flask by a siphoning action. This process may be continued automatically for effective separation of the desired components, which will then be collected into the solvent in the boiling flask.

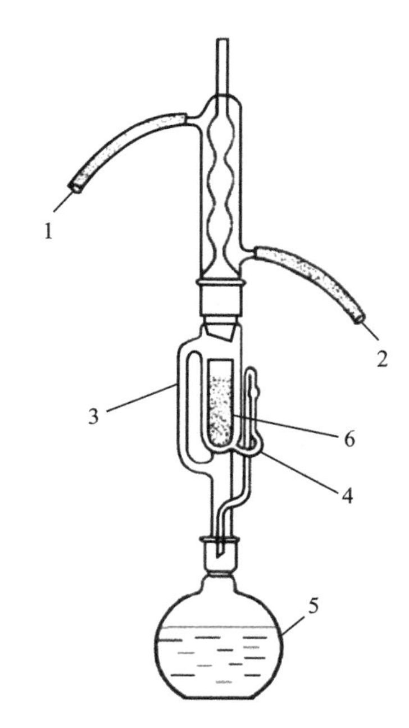

Figure 2-14 Soxhlet Extractor Apparatus

1—Water out; 2—Water in; 3—Vapor; 4—Siphon arm; 5—Boiling flask containing extracting solvent; 6—Porous thimble(to hold solid)

2. Sublimation for purification

Sublimation is a phase change in which a solid passes directly into the vapor phase without going through an intermediate liquid phase. Many solids having appreciable vapor pressures below their melting points can be purified by sublimation, either at atmospheric pressure or under vacuum. Such purification is most effective if the impurities have either low or very high vapor pressures at the sublimation temperature, so that they either do not sublime appreciably or fail to condense after vaporization. Although not as selective as recrystallization or chromatography, sublimation offers advantages in that no solvent is required, losses in transfer can be kept very low, and the process is rapid with the right apparatus and conditions. Sublimation can be accelerated by performing it under reduced pressure(vacuum sublimation), or in an air stream(entrained sublimation). A common sublimation apparatus is illustrated in Figure 2-15. As heating the

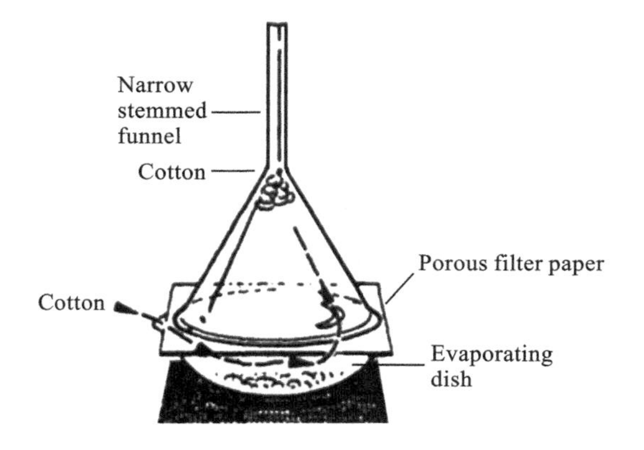

Figure 2-15 Apparatus for Simple Sublimation

evaporation dish from the bottom under the melting point of caffeine, caffeine sublimes and gas goes up through the porous filter paper so that they condense on the funnel or the paper.

Ⅲ Apparatus、Materials and Reagents

【Apparatus】 Scheme 1: Soxhlet extractor, boiling flask, Allihn, condenser, Liebig condenser, thermometer, evaporation dish, narrow-stem funnel.

Scheme 2: 500 mL beaker, 500 mL separatory funnel, 50 mL round-bottom flask, simple distillation apparatus.

【Materials】 Cotton, round filter paper sheet.

【Reagents】 Scheme 1: tea leaves, 2 g, 95% ethanol; calcium oxide, 4 g.

Scheme 2: tea leaves, 25 g; sodium carbonate, 20 g; dichloromethane, 50 mL; anhydrous $MgSO_4$; acetone; petroleum ether (60～90 ℃).

【Physical constants】

Compound	m. p. /℃	Sublimation Point/℃	Solubility/[g · (100 mL)$^{-1}$]			
			Water	Ethanol	Acetone	Dichloromethane
Caffeine	235～238	178	2.2(25 ℃) 18.2(80 ℃) 66.7(100 ℃)	1.5(25 ℃) 4.5(60 ℃)	2(25 ℃)	18.2(25 ℃)

Ⅳ Procedures

Scheme 1 Soxhlet Extraction

1. Extraction

Weigh 2 g of dry tea leaves and finely grind them, then pack them into a filter paper thimble and place the thimble into a Soxhlet extractor. Assemble the Soxhlet extractor apparatus(see Figure 2-14). Place several boiling stones in the distillation flask before attaching the rest of the glassware. Add 95% ethanol cautiously from the top of the Soxhlet extractor until it starts a siphoning action and then add 10 mL of more ethanol. Attach the condenser on the top and heat the flask to reflux for about 1 h(the siphoning action may occur 3～4 times). Stop heating after the last siphoning when liquid turn slightly green.

2. Distillation

After the apparatus cool down, disassemble the condenser and Soxhlet extractor, then place two boiling stones into the flask and assemble the apparatus for a simple distillation to remove the ethanol (See "Simple distillation" section). Stop heating when 10～15 mL of dark residues are left in the flask.

3. Sublimation

Transfer all the dark residues into an evaporation dish, place it on the hot plate with an asbestos center gauze on it for avoiding straight heating, hold the dish with a clamp and gently heat to evaporate the solvent until the solution turns thick, then add 4 g of CaO powder(about 2 spoons)and stir rapidly to remove those organic acids or water completely. Heating should be very cautious here, otherwise the caffeine would sublime and escape now before you collect it! Spread the dried green powder of solid residues on the bottom of dish, cover a porous filter paper and an inverted funnel which can cover the dish properly(see Figure 2-15). Control the temperature at about 200 ℃ for 20 min. After a few crystals are observed on the funnel, stop heating and transfer the sublimation apparatus and the asbestos center gauze together carefully onto the table. After it cools down, remove the funnel and scrape the sublimed caffeine recovered. Weigh your products and calculate the weight percentage recovery of caffeine from tea leaves.

Scheme 2　Leaching extraction

Dissolve 20 g of $NaCO_3$ into 250 mL of distilled water in a 500 mL beaker. Weigh 25 g of tea leaves packed with bundle, place the package into the $NaCO_3$ solution prepared above, and carefully heat the solution for boiling for 0. 5 h. Avoid foaming during the heating. Then take out the tea bag, press to dry and filter the solution by a vacuum filtration while it is hot. When the filtrate is cold, extract it with 50 mL CH_2Cl_2 in a 500 mL separatory funnel. To avoid emulsions, shake the solution to be extracted very gently until you see that two layers will separate readily. Then prepare a funnel with a small wad of cotton stemmed in the neck, and place 1 cm high anhydrous $MgSO_4$ on the cotton. Let out the lower organic layer directly from the separatory funnel to the $MgSO_4$ covered funnel and collect the filtrate into a Erlenmeyer flask. Wash the drying agent by 2～3 mL CH_2Cl_2. The left aqueous layer in the separatory funnel is extracted by another 50 mL CH_2Cl_2. Again let out the lower organic layer to be filtrated through some new $MgSO_4$. Then combine the filtrate from the two times of filtration. If the organic liquid is not clear or there are some water droplets mixed in it, another extraction and drying should be carried out in the same way.

Transfer the dry organic layer to a 50 mL round-bottom flask, add several boiling stones, and distill to retrieve the solvent CH_2Cl_2 under reflux on a water bath. Pump out the solvent thoroughly by a water aspirator. The residue, which contains caffeine, is purified by recrystallization with acetone-petroleum mixed solvent. The procedure is: Dissolve the residue in a small amount of acetone, then add petroleum(60～90 ℃)slowly until the solution is slightly turbid and let the product crystallize at 25 ℃ and then at 5 ℃. Collect the well-formed crystals after a vacuum filtration. Dry the crystals, then weigh them and calculate the weight percentage recovery of caffeine from tea leaves.

Ⅴ Notes and Instructions

【Requirements for preview】

(1) Get to learn about the principle of extraction and the application of extraction in the medical and pharmaceutical field.

(2) Review how to assemble the reflux and simple distillation apparatus.

【Experimental precautions】

(1) The siphoning arm of the extractor is very easily broken if it is held too tightly. Take care when you assemble the Soxhlet apparatus.

(2) Finely grinded tea leaves can help improve the efficiency of extraction, but be sure to pack them finely to avoid those spilling powder blocking the siphoning arm.

(3) Don't forget adding boiling stones before you start a simple distillation.

(4) If the left residue liquid is too thick after an over-exercised distillation, you can pour back a few amount of ethanol you've just recovered from distillation. Rinse the residue and transfer it into the evaporation dish as completely as possible.

(5) Pour the recovered ethanol into the designated container and discard the waste tea leaves residues into waste bin after the experiment finished.

Ⅵ Post-lab Questions

(1) The Soxhlet apparatus must be set up upright and not slant. Give an explanation.

(2) What are advantages and disadvantages of sublimation for purification of solids?

Ⅶ Verbs

Soxhlet extractor 索氏提取器;
siphoning arm 虹吸管;
evaporation dish 蒸发皿;
sublimation 升华;
sublimate 升华物

实验十　减压蒸馏

一、实验目的

(1) 了解减压蒸馏的基本原理。

(2) 掌握减压蒸馏的装置和操作。

二、实验原理

很多有机化合物,尤其是高沸点的有机化合物,在常压下蒸馏时往往部分或全部分解。在这种情况下,采用减压蒸馏的方法最为有效。一般的高沸点有机化合物,当压力降低至0.0027 MPa(20 mmHg)时,其沸点要比常压下的沸点低100～120 ℃。

由蒸馏的原理可知,物质的沸点随外界压力的降低而降低。根据热力学原理,在给定压力下的沸点可近似地由下述公式求出:

$$\lg p = A + B/T$$

式中,p 为蒸气压,T 为沸点(热力学温度),A、B 为常数。

如以 $\lg p$ 为纵坐标,$1/T$ 为横坐标作图,可以近似地得到一条直线。因此,可以从两组已知的压力和温度算出 A 和 B 的数值。再将所选择的压力代入上式,即可算出液体的沸点。另外,还可从图2-16所示的沸点-压力经验关系(根据国际标准单位制,压力的单位应为Pa,1 mmHg=0.133 kPa)近似地推算出高沸点物质在不同压力下的沸点。

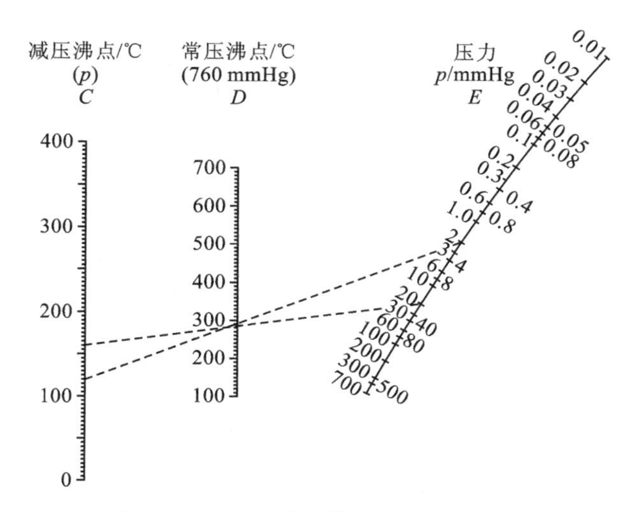

图 2-16　有机液体的沸点-压力经验关系

例如，某物质在常压下的沸点为 290 ℃，减压蒸馏时，若体系压力为 20 mmHg(2.66 kPa)，用尺子连接 E 上的 20 mmHg(2.66 kPa)与 D 上的 290 ℃两点，延长至 C 上的 160 ℃，即可得到该物质在 20 mmHg 下的沸点(约为 160 ℃)，表示为 160 ℃/2.66 kPa。反之同理。有些化学手册则直接给出化合物的蒸气压与沸点关系图(见图 2-17)。

典型的减压蒸馏装置如图 2-18 所示，整个系统由蒸馏、抽气(减压)、安全保护及测压装置四部分组成。待蒸馏液体的体积不能超过蒸馏瓶容积的 1/2，选择油浴或其他合适的方式加热，蒸馏液的液面应低于油浴液面。减压蒸馏时严禁使用有裂纹或薄壁玻璃仪器，防止爆裂；严禁使用非耐压容器，如平底锥形瓶，防止内向爆炸。

蒸馏装置的克氏蒸馏头的上口插末端拉成毛细管的厚壁玻璃管，毛细管下端插入液体且贴近圆底烧瓶底部(离瓶底 1～2 mm)，上端连有一段带螺旋夹的橡皮管。螺旋夹用以调节进入空气的量，使极少量的空气进入液体，呈微小气泡冒出，作为液体沸腾的汽化中心，使蒸馏平稳进行，同时起搅拌作用，因此减压蒸馏装置加热时不需要使用沸石。减压蒸馏装置中通常用蒸馏烧瓶等耐压容器作为接收器，并用耐压的厚壁橡皮管通过真空尾接管将其与作为缓冲用的抽滤瓶(安全瓶)连接起来。若待蒸馏液有多个组分需要收集，需用多尾真空尾接管连接多个接收器，收集不同馏分时，只需转动尾接管即可，而不用改变系统的真空度。此时，应在尾接管与冷凝管的接口处涂抹适量的真空脂，在保持气密性的同时方便尾接管在减压下转动。安全瓶的作用则是使减压蒸馏装置内的压力不发生突然变化以及防止减压装置中的水或泵油倒吸。

减压装置可用水泵或油泵，水泵可把压力降至 1.995～2.66 kPa(15～20 mmHg)，这对一般的减压蒸馏已经足够了。油泵可将压力降至 0.266～0.532 kPa(2～4 mmHg)。使用油泵时，需注意防护保养，不能使有机物、水、酸等的蒸气进入泵内。因此，用油泵进行减压蒸馏时，应在缓冲瓶和油泵之间依次安装冷阱、水银压力计和干燥塔。冷阱中常用冰-水、冰-盐或干冰作为冷却剂，以冷却水蒸气和一些易挥发的有机溶剂；干燥塔则依次装有无水氯化钙或硅胶(吸收水蒸气)、粒状氢氧化钠(吸收酸性气体)和切片石蜡(吸收烃类气体)。

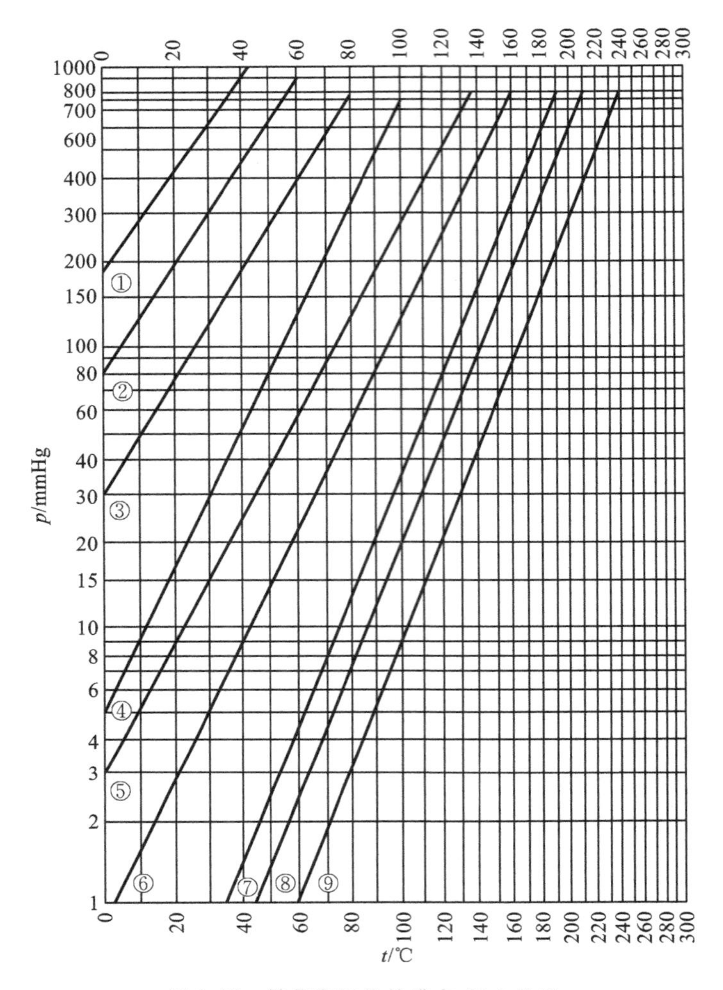

图 2-17　某些有机物的沸点-压力关系

1—乙醚；2—丙酮；3—苯；4—水；5—氯苯；6—溴苯；7—苯胺；8—硝基苯；9—喹啉

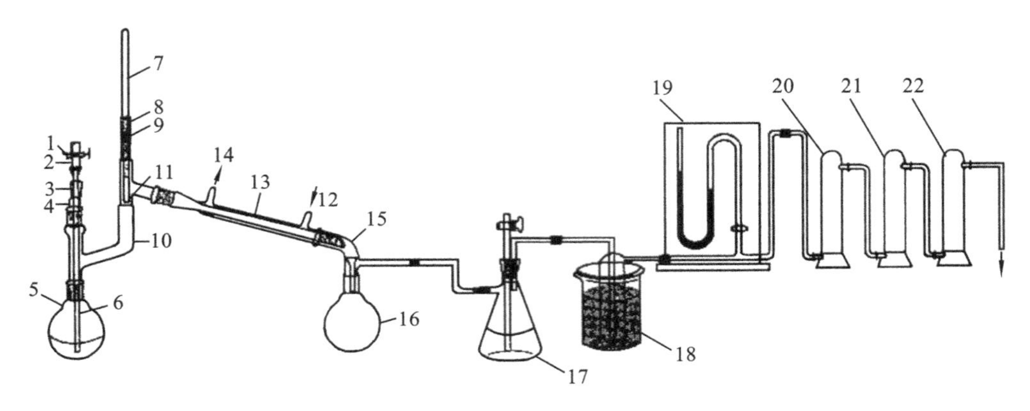

图 2-18　减压蒸馏装置

1—螺旋夹；2—乳胶管；3,8—单孔塞；4,9—套管；5—圆底烧瓶；6—毛细管；7—温度计；10—克氏蒸馏头；11—水银球；12—进水；13—直形冷凝管；14—出水；15—真空尾接管；16—接收瓶；17—安全瓶；18—冷阱；19—水银压力计；20—氯化钙塔；21—氢氧化钠塔；22—石蜡塔测压装置一般为封闭式水银压力计。为避免污染，仅在读取系统压力时将压力计与蒸馏系统连通。此时，所读的水银柱高度差(mm)即为系统压力(mmHg)。

减压蒸馏的主要操作步骤如下：

(1) 检查抽气泵的效率，真空度应满足要求。

(2) 按图 2-18 安装仪器，磨口处涂真空脂以增加气密性使用双叉尾接管。

(3) 开动油泵。

(4) 拧紧螺旋夹，直至橡皮管几乎封闭。

(5) 缓慢关闭安全瓶上的活塞。

(6) 几分钟后，记录压力。如压力不符合要求，检查所有接口处是否严密。获得良好的真空度后，才能继续后续操作。

(7) 缓慢打开活塞，待内外压力逐渐平衡后，关闭油泵，解除真空。

(8) 加入待蒸馏液，确保毛细管接近圆底烧瓶底部。打开油泵，并调节解压阀至所需压力。

(9) 观察毛细管冒泡情况，调节螺旋夹，活塞关闭时，应有连续小气泡通过液体。

(10) 调节升降台升高热源，开始加热。

(11) 当冷凝的蒸气环上升至温度计水银球且温度已恒定后，蒸馏开始，记录蒸馏时的温度、压力，蒸馏速度应保持在每秒 1 滴左右。

(12) 当新馏分蒸出时，旋转双叉尾接管更换接收瓶收集相应馏分。

(13) 蒸馏结束，移去热源，让蒸馏瓶冷却，慢慢旋开螺旋夹(防止倒吸)，并缓慢打开二通活塞，平衡内外压力，使水银压力计的水银柱慢慢地恢复原状(若打开太快，水银柱快速上升，有冲破水银压力计的可能)。关掉油泵，移去接收瓶，并拆下玻璃仪器清洗。

三、仪器与试剂

【仪器】真空泵系统，减压蒸馏装置(厚壁烧瓶，克氏蒸馏头，直形冷凝管，双叉尾接管，单口烧瓶)，厚壁玻璃管，弹簧夹，冷阱。

【试剂】乙酰乙酸乙酯，苯甲醛。

【物理常数】

试剂	相对分子质量	密度/(g·cm^{-3})	熔点/℃	沸点/℃ (0.1 MPa)	溶解性	
					H_2O	乙醇
乙酰乙酸乙酯	130.14	1.03	−45	180.4	溶	溶
苯甲醛	106.12	1.04	−26	179.6	微溶	溶

四、实验步骤

方案一　乙酰乙酸乙酯的减压蒸馏

量取 15 mL 乙酰乙酸乙酯，加至 25 mL 的圆底烧瓶中，按图 2-18 安装好减压蒸馏装置。选择一个合适的毛细管代替沸石提供气泡中心，并将冰-盐混合物(冰与盐质量比为 100∶33)装入冷阱中[1]。然后仔细检查真空系统、蒸馏系统是否正常。开启油泵，关闭活塞开始抽真空，待真空度在 10～20 mmHg 范围内稳定后[2]，开始加热进行减压蒸馏。除去部分前馏分后[3]，收集78 ℃(18 mmHg)的馏分。当圆底烧瓶还剩少量残液时，停止加热，慢慢开启活塞通大气，稍后关闭油泵。(乙酰乙酸乙酯沸点的文献数据：100 ℃/80 mmHg；92 ℃/40 mmHg；82 ℃/20 mmHg；74 ℃/14 mmHg；67 ℃/10 mmHg；54 ℃/5 mmHg。)

记录收集的乙酰乙酸乙酯体积，计算收率。

方案二　苯甲醛的减压蒸馏

取15 mL苯甲醛，加至25 mL的圆底烧瓶中，安装好毛细管及减压蒸馏装置。如上述方法，收集101～103 ℃/12 mmHg馏分。(苯甲醛沸点的文献数据：126 ℃/40 mmHg；115 ℃/20 mmHg；105 ℃/14 mmHg；101 ℃/12 mmHg；95 ℃/10 mmHg。)

五、注解和实验指导

【注解】

[1] 冰-盐体系的最低温度为－21 ℃。

[2] 毛细管的进气量必须合适，太大会使液体冲入冷凝管。若蒸馏时毛细管断了，需更换新的毛细管，待液体冷却后，方可重新进行减压蒸馏。

[3] 需要时，应先用水泵以除去其中少量的低沸点溶剂以保护油泵，再用油泵进行减压蒸馏。

【预习要求】

(1) 了解减压蒸馏的原理及应用范围。

(2) 熟悉减压蒸馏装置中各部分的名称及正确的连接顺序。

【操作注意事项】

(1) 严禁使用有裂痕或薄壁的玻璃仪器，特别是平底瓶，如锥形瓶等，以防爆裂。

(2) 系统真空度达到最大且稳定后才能开始加热。

(3) 维持油浴温度比馏出液沸腾温度高25～30 ℃，避免过热。

六、思考题

(1) 减压蒸馏时，可否用沸石代替毛细管？为什么？

(2) 估计下列化合物在多大压力下，80 ℃时可沸腾。

① 戊酸(b. p. 96 ℃/3.07 kPa)；

② 乙苯(b. p. 136 ℃/0.1 MPa)；

③ 溴苯(b. p. 156 ℃/760 mmHg)。

Experiment 10　Vacuum Distillation

Ⅰ Objectives

(1) To understand the basic principle of vacuum distillation;

(2) To learn how to operate the vacuum distillation.

Ⅱ Principle

Many substances cannot be distilled at atmospheric pressure, either because they boil at such high temperatures that decomposition occurs or because they are sensitive to oxidation. In such cases purification can be accomplished by distillation at reduced pressure. This operation is called as vacuum distillation. For organic compounds with high boiling points, their boiling points at the pressure of 0.0027 MPa(20 mmHg) are generally 100～120 ℃ lower than those at atmospheric pressure.

On the basis of the principle of simple distillation, the boiling point (b. p.) decreases with the decrease in external pressure. Therefore the boiling point at a certain pressure can be approximately given by the following equation:

$$\lg p = A + B/T$$

Where p is pressure, T is boiling point, A and B are constants, respectively.

If $\lg p$ is plotted as y-axis and $1/T$ as x-axis, an approximate straight line can be obtained. Therefore A and B can be calculated from two series of known p-T values, on the basis of which the boiling point at any given pressure can be estimated. It can also be estimated from an empirical boiling point-pressure nomograph as shown in Figure 2-16 (according to the national standard, Pa is the unit of pressure, 1 mmHg=0.133 kPa).

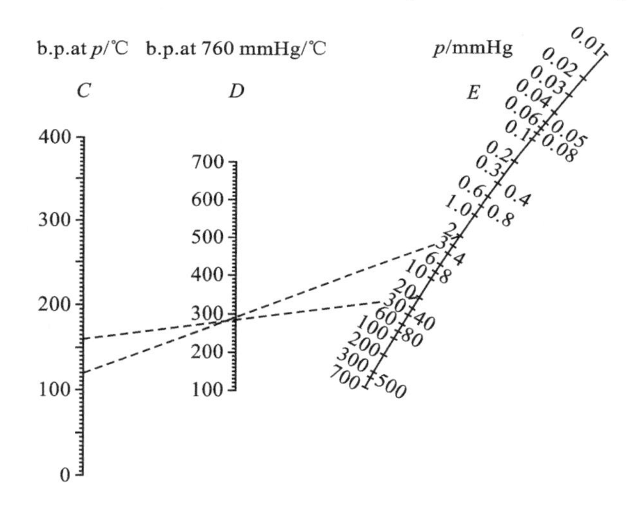

Figure 2-16　Empirical Boiling Point-pressure Nomograph for Organic Liquids

For example, the boiling point of a liquid is 290 ℃ at atmospheric pressure. When the pressure is reduced to 20 mmHg(2.66 kPa), we just simply place a ruler on the central line at the atmospheric boiling point of the compound (290 ℃), pivot it to line up with the appropriate pressure(20 mmHg) marking on the right-hand line, and read off the predicted boiling point from the left-hand line, which can be indicated as 160 ℃/20 mmHg. Similarly, we can estimate the boiling point at atmospheric pressure(about 295 ℃) on the central line according to the boiling point at reduced pressure(120 ℃/2 mmHg).

Some organic chemistry handbooks provide the values of A and B of some organic compounds. Thus the boiling point at any given pressure can be calculated directly according to the equation mentioned above. And some others show the relationship between boiling point and pressure for some common organic compounds (see Figure 2-17), in which the boiling point can also be obtained.

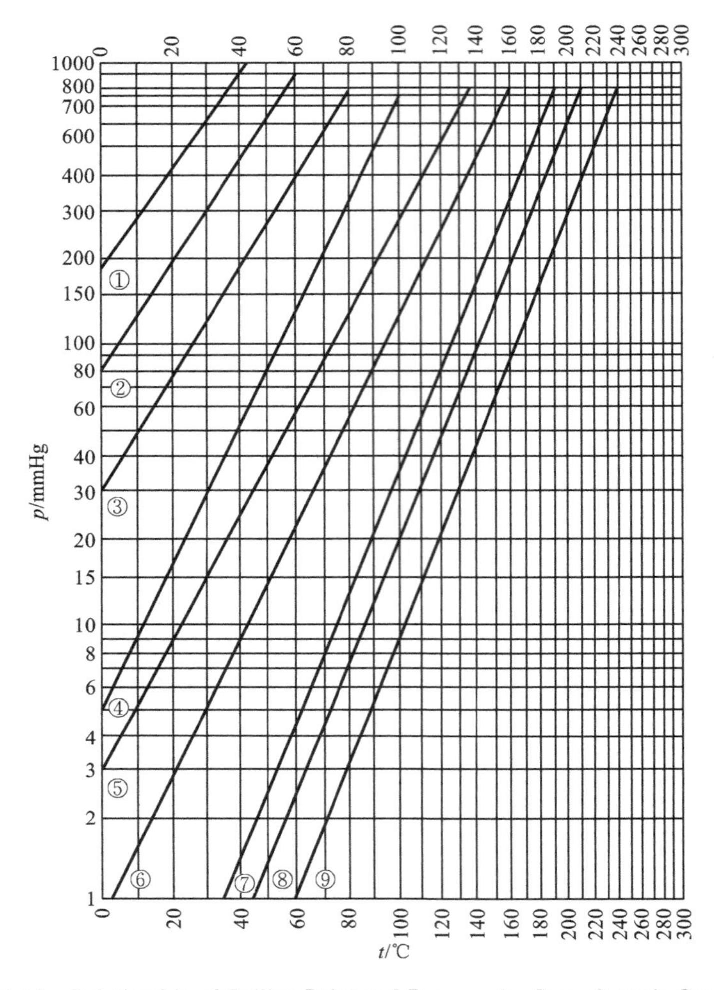

Figure 2-17 Relationship of Boiling Point and Pressure for Some Organic Compounds

1—ethyl ether; 2—acetone; 3—benzene; 4—water; 5—chlorobenzene;
6—bromobenzene; 7—aniline; 8—nitrobenzene; 9—quinoline

A typical apparatus for vacuum distillation are illustrated in Figure 2-18, which consists of four main parts, such as distillation, decompressor, safety protection device and manometrical system. An appropriate heating method should be selected according to the principle described in Experiment 1. The volume of the liquid to be distilled should not exceed half volume of the round-bottom distillation flask. And the surface of the liquid in the flask should be below that of the heating medium. The crack or thin-wall glassware is strictly prohibited to be used in vacuum distillation, to avoid burst at reduced pressure; the non-pressure-resistance glassware (e. g. conical flask) is also not permitted, to prevent implosion.

The left-hand neck of Claisen distilling head carries a screw-cap adapter, through which is inserted a heavy-wall glass tube of appropriate diameter drawn out to a capillary at its lower end. The capillary should be kept 1～2 mm away from the bottom of the distillation flask. The glass tube carries at its upper end a short piece of pressure tubing and a screw

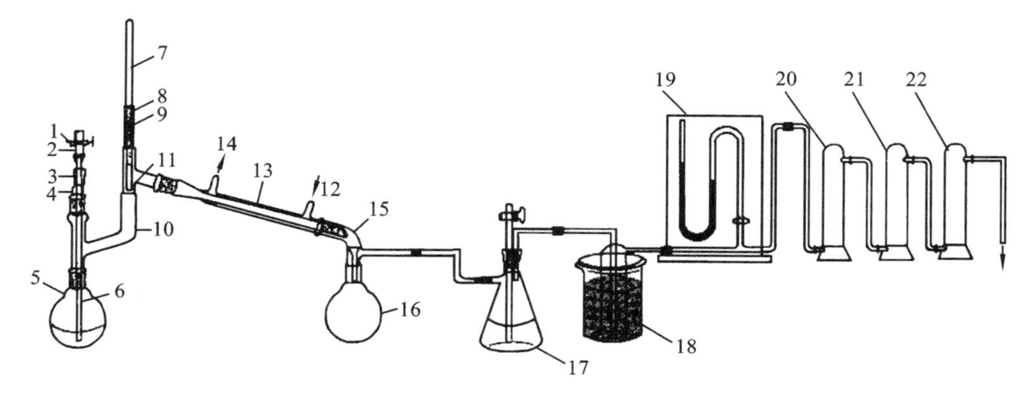

Figure 2-18 The Apparatus for Vacuum Distillation

1—screw clamp; 2—latex tube; 3,8—one pore stuff; 4,9—outlet adapter; 5—round-bottom flask; 6—capillary tube; 7—thermometer; 10—Claisen distilling head; 11—mercury bulb; 12—water in; 13—Liebig condenser; 14—water out; 15—vacuum adapter; 16—receiver; 17—surge flask; 18—cold trap; 19—mercury manometer; 20—$CaCl_2$ tower; 21—NaOH tower; 22—paraffin tower

clip. To adjust the screw clip can introduce very small amount of air into the liquid and produce steady bubbles to promote even boiling. The pressure-resistance vessel (e. g. distillation flask) is generally used as receiver to collect the distillate, which is connected to the surge flask using heavy-wall rubber tubing. If there are several distillates to be collected, a two or three-limbed multiple adapter, frequently called a "pig", will be used, which permits the collection of three individual fractions without breaking the vacuum and interrupting the progress of the distillation. The joint between adapter and condenser needs lubrication with appropriate amount of vacuum grease, which allows facile rotation of the adapter. The surge flask is fitted with a two-way Rotaflo stopcock to regulate the pressure in the distillation, which is essential since a sudden fall in pressure may result in the water or oil being sucked back.

Both water aspirator and oil pump can be used to produce vacuum in the distillation. When using water aspirator, it can reduce the pressure to 1.995～2.66 kPa(15～20 mmHg), which is sufficient in most cases. Oil pump can reduce the pressure as low as 0.266～0.532 kPa(2～4 mmHg). To protect the oil pump from the vapors of organic compounds, water and acid, the cold trap, mercury manometer and several drying towers should be assembled successively between the surge flask and pump. In the cold trap, ice-water, ice-salt or dry ice is commonly used as coolant to cool down the water vapor and volatile organic solvent. The drying towers are loaded with anhydrous calcium chloride or silica gel (to absorb the water vapor), granulous sodium hydroxide (to absorb acidic gas) and scale paraffin (to absorb the vapor of hydrocarbons) successively.

The pressure of the system is measured with a closed-end mercury manometer. To avoid contamination of the manometer, it should be connected to the system only when a reading is being made. The pressure, in mmHg, is given by the height difference of two mercury columns in mm.

The general process of vacuum distillation is described as follows:

(1) Choose an appropriate pump that meets the requirement of vacuum degree.

(2) Assemble the apparatus according to Figure 2-18, vacuum grease can be smeared to each joint to ensure air tightness.

(3) Start the oil pump.

(4) Tighten the screw clip until the rubber tubing is nearly sealed.

(5) Turn off the stopcock of the surge flask slowly.

(6) Measure the pressure after a few minutes. If the pressure isn't low enough, check all joints until the required vacuum degree is achieved.

(7) Turn on the stopcock slowly to release vacuum, and then turn off the oil pump.

(8) Add the liquid to the round-bottom flask, and make sure the capillary tube is near the bottom of the flask. Turn on the oil pump, and then turn off the stopcock of surge flask slowly until the system pressure is stable.

(9) Observe the bubbles from capillary tube, and adjust the screw clip to maintain continuous small bubbles getting through the liquid.

(10) Adjust the lifting platform to heat the distillation flask.

(11) Keep distilling rate at 1 drop per second after the vacuum and temperature are stable.

(12) If more fractions are to be collected, just rotate the fraction collector slowly and carefully to change a new receiver without breaking the vacuum system.

(13) When distillation is finished, remove the heating source firstly. Only when the distillation flask is cooling, one can release the vacuum, and then turn off the oil pump. Disassemble the apparatus step by step and rinse all the glassware.

Ⅲ Apparatus and Reagents

【Apparatus】 Vacuum pump, vacuum distillation apparatus, thermometer, spring clip.

【Reagents】 Ethyl acetoacetate(15 mL), benzaldehyde(15 mL).

【Physical constants】

Reagent	M_w	ρ/(g · cm^{-3})	m. p. /℃	b. p. /℃ (at 0.1 MPa)	Solubility	
					H_2O	Ethanol
Ethyl acetoacetate	130.14	1.03	−45	180.4	Soluble	Soluble
Benzaldehyde	106.12	1.04	−26	179.6	Slightly soluble	Soluble

Ⅳ Procedures

Scheme 1 Vacuum Distillation of Ethyl Acetoacetate

Add 15 mL of ethyl acetoacetate into a 25 mL round-bottom flask. Assemble the vacuum distillation apparatus according to Figure 2-18. Choose a suitable capillary tube instead of boiling stones to produce bubbles. Fill the cold trap with ice-salt mixture(100 : 33)[1], and check the vacuum and distillation system carefully to ensure it work well. Start the oil pump and shut down the stopcock. When the pressure is stable in the range of 10～20 mmHg[2], heat the round-bottom flask. After removing the fore-run fraction, collect the fraction at 78 ℃(18 mmHg)[3]. Stop heating when there is only a small amount of residual liquid left. Open the stopcock slowly to release vacuum, and then turn off the oil pump. (The reported boiling points of ethyl acetoacetate in literature: 100 ℃/80 mmHg; 92 ℃/40 mmHg; 82 ℃/20 mmHg; 74 ℃/14 mmHg; 67 ℃/10 mmHg; 54 ℃/5 mmHg.)

Write down the volume of ethyl acetoacetate and calculate the recovery rate.

Scheme 2　Vacuum Distillation of Benzaldehyde

Pour 15 mL of benzaldehyde to a 25 mL round-bottom flask. The following steps are the same to those described above. Collect the fraction at 101～103 ℃/12 mmHg. (The reported boiling points of benzaldehyde in literature: 126 ℃/40 mmHg; 115 ℃/20 mmHg; 105 ℃/14 mmHg; 101 ℃/12 mmHg; 95 ℃/10 mmHg.)

Ⅴ Notes and Instructions

【Notes】

[1] The lowest temperature that ice-salt mixture(100 : 33)can reach is －21 ℃.

[2] The inflow rate of air should be appropriate; otherwise, too much air would cause bumping and push the liquid into the condenser. If capillary tube is broken, change a new one and one cannot start the vacuum distillation immediately before the liquid is cooled down.

[3] If needed, water aspirator should be firstly used to remove the fraction with low boiling point, and then the oil pump.

【Requirements for preview】

(1) To understand the principle of vacuum distillation and its application.

(2) To be familiar with all the apparatus for vacuum distillation and the right order to assemble it.

【Experimental precautions】

(1) Never use the cracked, thin-wall or flat-bottom glassware (e. g. conical flask) in vacuum distillation.

(2) Heating is strictly prohibited before the required vacuum degree is obtained and becomes stable.

(3) In vacuum distillation, traditional boiling stones can't work, so capillary tube should be used to produce bubbles, that prevent bumping of the liquid at reduced pressure.

(4) When a boiling point is recorded, the corresponding pressure should also be written down at the same time. For example, the boiling point of benzaldehyde is 180 ℃ at atmospheric pressure and 87 ℃ at 35 mmHg. Thus it can be recorded as b. p. 180 ℃/760 mmHg and 87 ℃/35 mmHg, respectively.

(5) The temperature of oil bath or other heating source should be 25～30 ℃ above the boiling point of the liquid at the recorded pressure.

(6) To protect the oil pump, water aspirator should be firstly used to remove the component with low boiling point.

Ⅵ Post-lab Questions

(1) Can boiling stones be used in vacuum distillation instead of capillary tube? Why?

(2) Estimate the pressures that the following compounds can boil at 80 ℃.

① Pentanoic acid (b. p. 96 ℃/3.07 kPa);

② Ethylbenzene (b. p. 136 ℃/0.1 MPa);

③ Bromobenzene(b. p. 156 ℃/760 mmHg).

Ⅶ Verbs

vaccum distillation 减压蒸馏;

vacuum degree 真空度;

oil pump 油泵;

cold trap 冷阱;

drying tower 干燥塔;

mercury manometer 水银气压计

实验十一 水蒸气蒸馏

一、实验目的

(1) 掌握水蒸气蒸馏的装置及其操作方法。

(2) 了解水蒸气蒸馏的原理及其应用。

二、实验原理

不溶或几乎不溶于水的挥发性有机物可以用水蒸气蒸馏法通过水和有机物共沸来达到分离提纯的目的。当各物质不相溶时,各组分的分压(p_i)与纯物质的蒸气压(p_i°)相等,即

$$p_i = p_i^\circ \tag{2-1}$$

根据道尔顿原理,不相溶混合物的蒸气压等于各组分蒸气压之和,即

$$p_T = p_a^\circ + p_b^\circ + \cdots + p_i^\circ \tag{2-2}$$

从式(2-2)可知,由于其他组分蒸气压的作用,混合物总压力比最易挥发组分的蒸气压大,混合物沸腾温度一定比组分中最低沸点还低。例如,水(b. p. 100 ℃)和溴苯(b. p. 156 ℃)互不相溶,其纯物质及混合物的蒸气压－温度曲线如图 2-19 所示。

由图 2-19 可知,95 ℃时,混合物的蒸气压等于外压,混合物应在 95 ℃左右沸腾。这与理论预测一致,该温度低于水的沸点。由于水蒸气蒸馏的温度低于 100 ℃,因而具有广泛用途,特别适用于对热敏感,高温会分解物质的纯化。在有机反应中,常有焦油状物生成,对从焦油状混合物中分离出有机物,水蒸气蒸馏十分有用。

水蒸气蒸馏液的组成与化合物的摩尔质量及蒸馏温度下各蒸气压有关。

对于二组分混合物 A 和 B,如果 A 和 B 蒸气近似理想气体,可应用理想气体方程,得到下面的表达式:

$$p_A^\circ V_A = (m_A/M_A)RT, \quad p_B^\circ V_B = (m_B/M_B)RT \tag{2-3}$$

式中,p_A°、p_B° 为纯液体蒸气压,V_A、V_B 为气体体积,m_A、m_B 为气相组分质量,M_A、M_B 为摩尔质

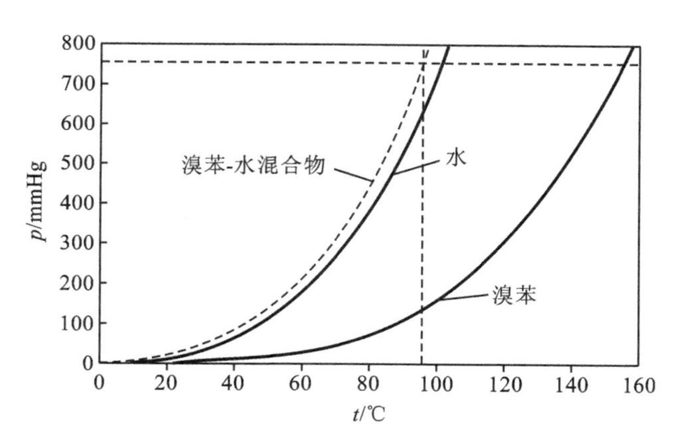

图 2-19　蒸气压-温度曲线

量，R 为摩尔气体常数，T 为热力学温度。两式相除得

$$\frac{p_{\mathrm{A}}^{\circ}V_{\mathrm{A}}}{p_{\mathrm{B}}^{\circ}V_{\mathrm{B}}}=\frac{m_{\mathrm{A}}M_{\mathrm{B}}RT}{m_{\mathrm{B}}M_{\mathrm{A}}RT} \tag{2-4}$$

因为分子、分母中的 RT 是相同的，气体体积相同($V_{\mathrm{A}}=V_{\mathrm{B}}$)，式(2-4)可变为

$$\frac{m_{\mathrm{A}}}{m_{\mathrm{B}}}=\frac{p_{\mathrm{A}}^{\circ}M_{\mathrm{A}}}{p_{\mathrm{B}}^{\circ}M_{\mathrm{B}}} \tag{2-5}$$

可通过溴苯的水蒸气蒸馏说明上述原理。混合物溴苯和水在 95 ℃时，蒸气压分别是 120 mmHg 和 640 mmHg(见图 2-19)，可用式(2-5)计算蒸馏液组成，即

$$\frac{m(\text{溴苯})}{m(\text{水})}=\frac{120\times157}{640\times18}=\frac{1.64}{1}$$

以质量计算，尽管在蒸馏温度下，溴苯的蒸气压比水低很多，但蒸馏液中溴苯的含量比水高。这是因为有机物摩尔质量通常比水大得多，只要在 100 ℃时有 5 mmHg 左右的蒸气压，就可以用水蒸气蒸馏得到良好效果的提纯，甚至固体也可以用水蒸气蒸馏提纯。

水蒸气蒸馏除需常压蒸馏的仪器以外，还需要水蒸气发生器和导管，另外需一长颈的圆底烧瓶，其装置如图 2-20 所示。

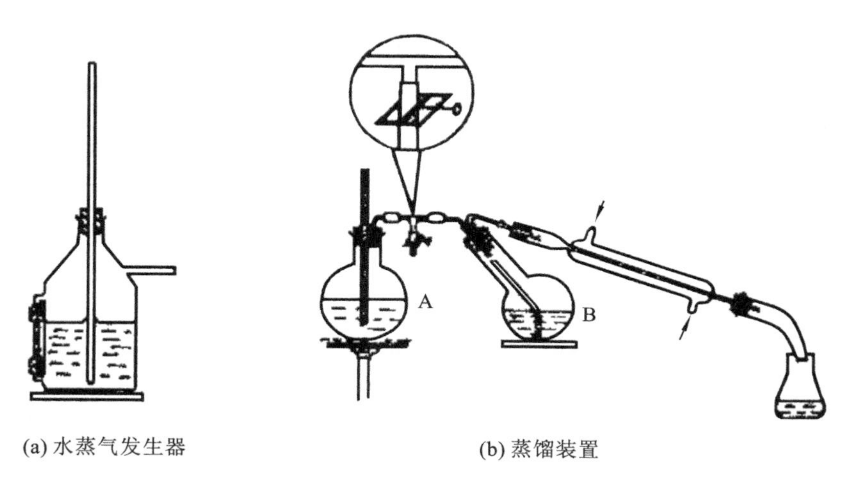

(a) 水蒸气发生器　　(b) 蒸馏装置

图 2-20　水蒸气蒸馏装置

水蒸气可由水蒸气发生器产生,水蒸气发生器是铁质或铜质的,也可以用圆底烧瓶代替,发生器内盛的水约占总容量的一半,从侧面的玻管可知发生器内水量的多少,中间的玻管称为安全管,管内水柱的高低反映内部蒸汽的压力。水蒸气发生器与蒸馏瓶之间有一导管,其下端有一螺旋夹。导管不要太长,塞子都要紧密,以防漏气。圆底烧瓶应用铁夹夹紧,使它斜置,与桌面成45°角,以避免跳溅的液沫被蒸气带进冷凝管。瓶口配置双孔塞,一孔插入水蒸气导气管,另一孔插入蒸气的出口导管。

三、仪器、试剂与材料

【仪器】水蒸气蒸馏装置。

【试剂与材料】硝基苯,萘,乙醚,橘子皮。

【物理常数】

试剂	相对分子质量	密度/(g·cm^{-3})	熔点/℃	沸点/℃	溶解度	
					H_2O	乙醇
硝基苯	123.11	1.205	5.7	210.9	不溶	溶
萘	128.18	1.162	80.5	217.9	不溶	溶

四、实验步骤

方案一　硝基苯的水蒸气蒸馏

按图2-20装好水蒸气蒸馏装置[1],取下长颈圆底烧瓶,放入10 mL硝基苯和40 mL纯水,再装上圆底烧瓶。检查各塞子是否严密后松开T形管的螺旋夹,加热水蒸气发生器使水沸腾。当有水蒸气从T形管冲出时,关紧螺旋夹,使水蒸气通入烧瓶中。为了使蒸气不致在烧瓶中冷凝,可在烧瓶底下垫石棉网,用小火加热。蒸馏过程中如果安全管内水柱上升从顶端喷出,说明蒸馏系统内压力增高,应立即打开螺旋夹,并停止加热。检查管道有无堵塞,排除故障。当馏出液澄清透明时,可停止蒸馏。停止蒸馏时应先松开螺旋夹,再移去热源,以防倒吸。

将全部蒸出液倒入分液漏斗中,先分出硝基苯,剩余的水层加入30 mL乙醚,振荡,放置分层后,将醚层与前边分出的硝基苯合并于干燥的小锥形瓶中,加入1.0～1.5 g黄豆粒大小的无水氯化钙,塞好瓶口,干燥0.5 h,其间应振荡几次。然后改用常压蒸馏装置,在温水浴中将乙醚蒸出,即可得硝基苯(蒸乙醚操作参见实验三)。

方案二　橘子皮的水蒸气蒸馏

按图2-20装置完毕后,将待蒸馏的橘子皮置于烧瓶中(占1/3左右)。其他步骤同方案一。蒸馏结束后将馏出液倾入指定的瓶中以供集中使用。

方案三　萘的水蒸气蒸馏

用克氏蒸馏头代替长颈圆底烧瓶进行水蒸气蒸馏(见图2-21)[2]。称取4.0 g粗品萘,加入50 mL圆底烧瓶中,蒸馏过程中冷凝管中的水要时开时停,随时注意,不要使蒸馏出的萘冷凝成固体后把尾接管堵死。也可以不加尾接管,冷凝管直接与接收瓶相连,待蒸出液透明后,再多蒸出10～15 mL清液。然后用抽滤的方法,收集产品,并测熔点。

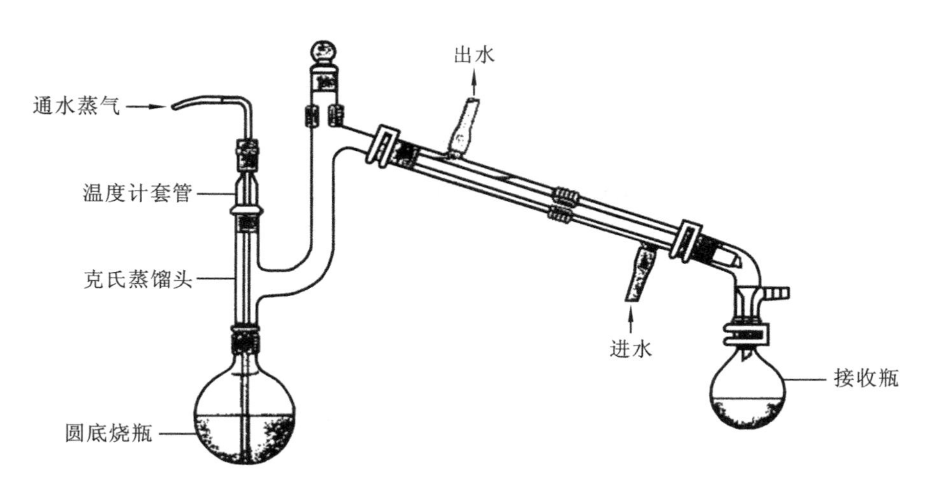

图 2-21　水蒸气蒸馏装置(用克氏蒸馏头代替长颈圆底烧瓶)

五、注解和实验指导

【注解】

[1] 为增加冷却效果,所用的冷凝管要比常压蒸馏用的冷凝管长。

[2] 用磨口圆底烧瓶上连接一个克氏蒸馏头代替长颈圆底烧瓶,可防止瓶内的液体因跳溅而冲入冷凝管内。

【预习要求】

(1) 了解本实验纯化硝基苯的原理及操作过程中的注意事项。

(2) 熟悉分液漏斗的使用和保养方法。

【操作注意事项】

(1) 水蒸气蒸馏时,玻璃仪器非常烫,操作时要小心。

(2) 水蒸气蒸馏结束,先打开 T 形管的螺旋夹。否则,蒸馏瓶中热液体会倒流进入反应器。

(3) 蒸馏瓶内液体的体积不能超过总容积的 1/3。

(4) 在整个水蒸气蒸馏过程中,要仔细观察水蒸气发生器侧管和安全管中的水位以及圆底烧瓶中通入水蒸气的情况,以及时排除故障和防止倒吸等现象。

(5) 水蒸气蒸馏挥发物时,冷凝液是混浊的,共馏物冷凝后分层,出现混浊,一旦馏出液澄清,结束蒸馏。

(6) 水蒸气蒸馏的物质可能在冷凝管中固化,仔细观察,避免形成大块结晶,阻塞冷凝管。如大块晶体聚集冷凝管,暂时关闭冷凝水,并放掉冷凝管中的水。热蒸气将熔化晶体,除去阻塞物。阻塞物一除,立即再通冷凝水。

六、思考题

(1) 指出下列各组混合物采用水蒸气蒸馏法进行分离正确与否。为什么?

① 甲醇,b. p. 65 ℃(760 mmHg)和水;(甲醇与水混溶)

② 对二氯苯,b. p. 174 ℃(760 mmHg)和水;(对二氯苯不溶于水)

③ 乙二醇($HOCH_2CH_2OH$),b. p. 197.3 ℃(760 mmHg)和水。(乙二醇与水混溶)

(2) 为什么水蒸气蒸馏温度永远低于 100 ℃?

(3) 应用水蒸气蒸馏的化合物必须具有哪些性质?

(4) 水蒸气蒸馏有哪些优点和缺点?

Experiment 11　Steam Distillation

Ⅰ Objectives

(1) To understand the principle of steam distillation and its application.

(2) To learn how to assemble the apparatus of steam distillation.

Ⅱ Principle

Steam distillation is widely used to separate the insoluble but volatile organic compounds from nonvolatile compounds in water solution by passing steam into the mixture of the compound to be purified and water. If the substances in one system are immiscible, the partial pressure will be equal to their pure vapor pressure(p_i°), as shown in Equation (2-1).

$$p_i = p_i^\circ \tag{2-1}$$

On the basis of Dalton's rule, the total pressure is equal to the sum of the partial pressure of all components, as shown in Equation (2-2).

$$p_T = p_a^\circ + p_b^\circ + \cdots + p_i^\circ \tag{2-2}$$

It can be indicated that the total pressure of the mixture is higher than the partial pressure of each component, even the most volatile one. Therefore the boiling point of mixture is lower than that of each pure component.

For example, water (b. p. 100 ℃/760 mmHg) and bromobenzene (b. p. 156 ℃) are immiscible. It can be seen from Figure 2-19 that the mixture can be boiling at 95 ℃, and this temperature is lower than 100 ℃. Some high-boiling substances decompose before the boiling point is reached and, if impure, cannot be purified by ordinary distillation. However, they can be freed from contaminating substances by steam distillation at a lower temperature.

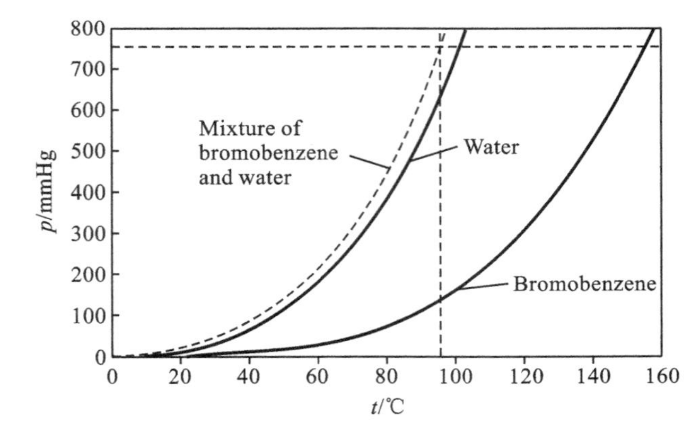

Figure 2-19　*p-t* Relationship of Bromobenzene-water System

The components of steam distillate are also related to the molecular weight of compounds and pressure at different temperature. For a binary mixture(A and B), if the behavior of gas A and B are similar to ideal gas, for which the equation can be used as:

$$p_A^\circ V_A = (m_A/M_A)RT,\quad p_B^\circ V_B = (m_B/M_B)RT \tag{2-3}$$

Where p_A°, p_B° is pressure of pure liquid, V_A, V_B is gas volume, m_A, m_B is gas weight, M_A, M_B is molecular weight, R is molecular gas constant and T is absolute temperature, respectively.

Equation (2-4) can be obtained from Equation (2-3), and it can be transformed to Equation (2-5) because of the equality of V_A and V_B

$$\frac{p_A^\circ V_A}{p_B^\circ V_B} = \frac{m_A M_B RT}{m_B M_A RT} \tag{2-4}$$

$$\frac{m_A}{m_B} = \frac{p_A^\circ M_A}{p_B^\circ M_B} \tag{2-5}$$

For the binary mixture of water (W) and bromobenzene (BB), the pressures of bromobenzene and water at 95 ℃ are 120 mmHg and 640 mmHg, respectively. So the composition of this mixture at 95 ℃ can be calculated according to Equation (2-5).

$$\frac{m_{BB}}{m_W} = \frac{120 \times 157}{640 \times 18} = \frac{1.64}{1}$$

It can be found that even the pressure of water is higher than that of bromobenzene, the weight of BB in distillate is larger than that of water because of the much higher molecular weight of BB. That means steam distillation can be used to purify organic compounds if only they possess 5 mmHg pressure at 100 ℃. Even solid organic compounds can also be purified by this method.

Besides all instruments used in simple distillation, steam generator, conduit and long-neck round-bottom flask are needed in steam distillation as shown in Figure 2-20.

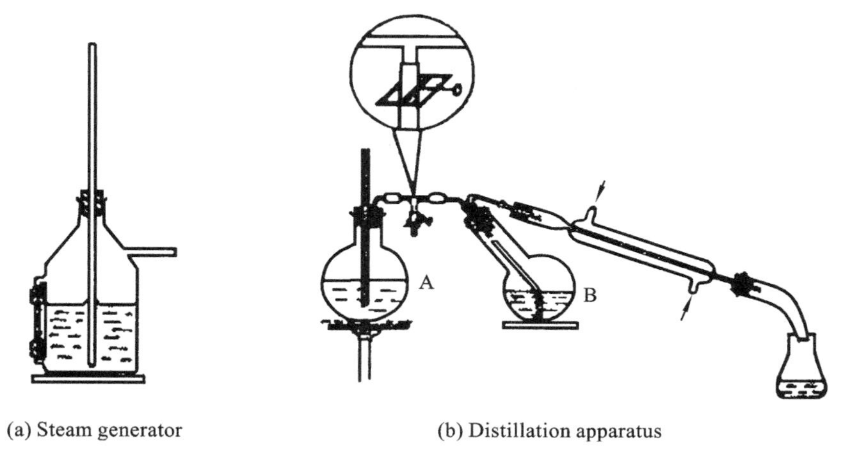

(a) Steam generator　　(b) Distillation apparatus

Figure 2-20　Steam Distillation Device

Steam can be produced by steam generator which is made of iron or copper. Sometimes steam generator can be replaced by a round-bottom flask, in which the water is heated for boiling to produce steam. The water should be less than half volume of generator and can be seen from the lateral glass pipe. The long glass pipe in the middle of generator is called safety pipe with the water column to show the steam pressure. A T-type conduit, locating between generator and distillation flask, is fitted with a screw clip at its bottom end. The conduit should not be too long and the plug should be tightened in case of leakage. To avoid hot bubbles being brought into condenser, the round-bottom flask should be fixed by iron clamp and 45° diagonal to table(as shown in Figure 2-20). Bottleneck should be stuffed by two stopple plug, of which one is interposed by steam conduit and the other is interposed by steam-out pipe.

Ⅲ Apparatus and Reagents

【Apparatus】 Steam distillation apparatus.

【Reagents】 Nitrobenzene (12.05 g, 10 mL, 0.098 mol), crude naphthalene (4.0 g), orange peel, ether.

【Physical constants】

Reagent	M_w	$\rho/(g \cdot cm^{-3})$	m. p. /℃	b. p. /℃	Solubility	
					H_2O	Ethanol
Nitrobenzene	123.11	1.205	5.7	210.9	Insoluble	Soluble
Naphthalene	128.18	1.162	80.5	217.9	Insoluble	Soluble

Ⅳ Procedures

Scheme 1　Steam Distillation of Nitrobenzene

Assemble the steam distillation apparatus according to Figure 2-20[1], and then add 10 mL of nitrobenzene and 40 mL of water to long-neck round-bottom flask. Check all junctions and loosen the screw clip, heat steam generator to boil the water, tighten the screw clip and pass the steam into the flask. To prevent steam from being condensed to liquid, the flask should also be warmed by gently heating. If water column in safety pipe sprays out, screw clip should be loosened quickly because of high pressure in system and stop heating. When the distillate becomes clear, the distillation can be ceased. Loosen the screw clip and move away the heating source.

Pour all distillate into a separatory funnel, let out the lower layer which contains nitrobenzene through the stopcock firstly. Then add 30 mL of ethyl ether to the residual water layer for exhausting extraction. After ethyl ether is separated, combine ethyl ether with nitrobenzene in a dry conical flask and add 1.0～1.5 g of anhydrous calcium chloride to dry the liquid. After standing for 0.5 h, remove the drying agents by filtration. Carry out simple distillation on a water bath to remove ethyl ether with nitrobenzene left in the flask.

Scheme 2　Steam Distillation of Orange Peel

Assemble the apparatus according to Figure 2-20, and 1/3 of the round-bottom flask is filled with orange peel. The following steps are the same as described in Scheme 1. After steam distillation, collect all distillate in specified flask.

Scheme 3 Steam Distillation of Naphthalene

Using a ground round-bottom flask with a Claisen distilling head to replace the long-neck round-bottom flask (see Figure 2-21)[2], and assemble the steam distillation apparatus according to Figure 2-21. Add 4.0 g of crude naphthalene to 50 mL round-bottom flask. Carry out steam distillation as described in Scheme 1. Keep on collecting more 10～15 mL of clear distillate when the distillate becomes transparent. After vacuum filtration, dry the solid and then measure its melting point.

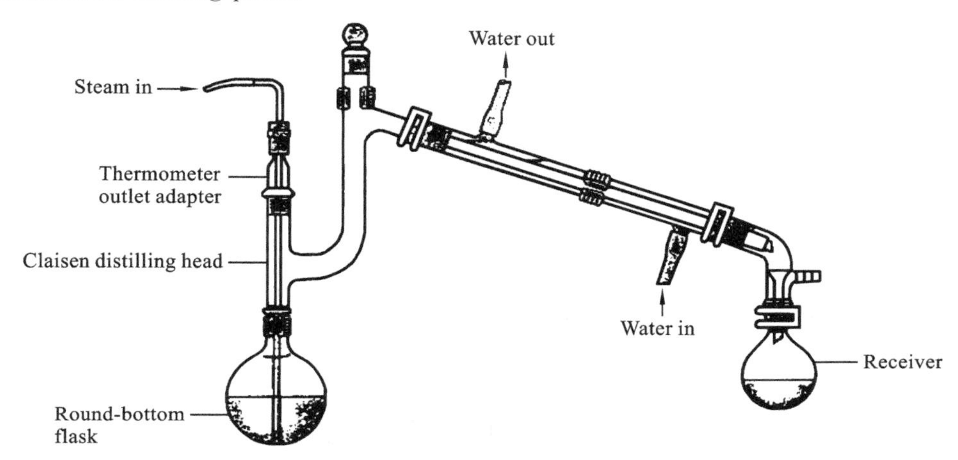

Figure 2-21 Steam Distillation Device (Long-neck of Round-bottom Flask Is Replaced by a Claisen Distilling Head)

Ⅴ Notes and Instructions

【Notes】

[1] To promote the cooling efficiency, the condenser should be a little longer than that used in simple distillation.

[2] Ground round-bottom flask with Claisen distilling head can be used instead of long-neck round-bottom flask, which can better avoid the boiling liquid splatter into the flask.

【Requirements for preview】

(1) To understand how to purify nitrobenzene by steam distillation.

(2) To be familiar with the usage of separatory funnel.

【Experimental precautions】

(1) Be careful of the hot glassware and steam.

(2) When steam distillation is over, loosen the T-type screw clamp firstly to prevent the backflow of hot liquid.

(3) The volume of the liquid to be steam distilled should be less than 1/3 of the vessel volume.

(4) Be careful of the water level in steam generator and steam in round-bottom flask during distillation to avoid any block and suck-back.

(5) The early distillate will be cloudy. Once it becomes clear, cease the distillation.

(6) Never allow the distillate form agglomerated crystals to block the condenser. Once it appears, temporarily stop condensate water. The crystals will be melted by hot steam, and then resume condensate water.

Ⅵ Post-lab Questions

(1) Whether can the following mixtures be separated by steam distillation or not? Why?

① Methanol (b. p. 65 ℃/760 mmHg) and water, methanol can dissolve in water;

② *p*-Dichlorobenzene (b. p. 174℃/760 mmHg) and water, *p*-dichlorobenzene can't dissolve in water;

③ Glycol($HOCH_2CH_2OH$, b. p. 197.3 ℃/760 mmHg) and water, glycol can dissolve in water.

(2) Why is the temperature of steam distillation always below 100 ℃?

(3) What characters should compounds to be steam distilled have?

(4) What are the advantages and disadvantages of steam distillation?

Ⅶ Verbs

steam distillation 水蒸气蒸馏;
Claisen distilling head 克氏蒸馏头
steam generator 水蒸气发生器;

实验十二 氨基酸的纸上电泳

一、实验目的

(1) 了解电泳的基本原理。

(2) 掌握纸上电泳分离、鉴定氨基酸的原理与方法。

二、实验原理

电泳是指在一定条件下带电质点在直流电场作用下,向着与其带相反电荷的电极方向移动的现象。由于混合物中各组分所带电荷性质、电荷数量以及相对分子质量、分子体积、分子形状等因素不同,在同一电场中,各物质泳动方向和速率都存在着差异。因此,在一定时间内,利用各组分的泳动距离不同而达到分离和鉴定的目的。电泳可分为显微电泳、自由界面电泳和区带电泳,其中区带电泳操作简便,分离效果好,最常用于分离鉴定。它是在不同的惰性支持物中进行的电泳,能使各级组分分成带状区间。以滤纸作为带电质点的惰性支持物进行的区带电泳称为纸上电泳。纸上电泳常用于分离氨基酸和蛋白质,也可用于无机离子、配位化合物、糖类、染料等物质。

一个带电质点在电场中的电泳速率除受本身性质的影响外,还受下列因素的影响:

(1) 电场强度。电场强度是指每厘米支持物的电位降,单位为 $V \cdot cm^{-1}$。它对电泳速率起着重要的作用。电场强度越高,带电质点移动速率越快。根据电场强度的大小,可将电泳分为高压电泳(大于 50 $V \cdot cm^{-1}$)、常压电泳(10~50 $V \cdot cm^{-1}$)和低压电泳(小于 10 $V \cdot cm^{-1}$)。

(2) 溶液的 pH 值。溶液的 pH 值决定了带电质点的解离程度,也决定了物质所带电荷的多少，对蛋白质、氨基酸等两性物质而言，溶液的 pH 值离等电点越远，质点所带净电荷越多，电泳速率越快。为了使电泳中支持介质保持稳定的 pH 值，电泳时必须使用缓冲溶液。

(3) 离子强度。电泳时溶液的离子强度越大，电泳速率越慢；如果离子强度太小，溶液的缓冲容量小，不易维持恒定的 pH 值。一般缓冲溶液的离子强度在 0.01～0.2。

(4) 电渗。电场中液体对固体支持物的相对移动称为电渗。它是由缓冲溶液的水分子和支持介质的表面之间所产生的一种相关电荷所引起的，如滤纸的孔隙带负电荷，与滤纸相接触的水溶液则带正电荷。若质点的电泳方向与电渗水溶液移动方向一致，则电泳速率加快；反之，电泳速率减慢。因此，应尽量选择电渗作用小的物质做支持物。

不同物质的电泳速率通常用离子迁移率来表示。其定义为带电质点在单位电场强度下的泳动速率，用公式表示为

$$\mu = \frac{u}{E} = \frac{d/t}{V/l} = \frac{dl}{Vt} \tag{2-6}$$

式中：μ 为离子迁移率，$cm^2 \cdot V^{-1} \cdot s^{-1}$；$u$ 为质点的泳动速率，$cm \cdot s^{-1}$；E 为电场强度，$V \cdot cm^{-1}$；d 为质点移动的距离，cm；l 为支持物的有效长度，cm；V 为加在支持物两端的实际电压，V；t 为电泳时间，s 或 min。

电泳所用的装置由电泳仪和电泳槽两个部分组成。不同厂商生产的各种电泳仪和电泳槽规格、样式很多，其工作原理相同，此处以 DY-1 型电泳仪为例说明。其装置包括电源装置和电泳槽两大部分。电源装置如图 2-22 所示，其内部有稳压整流装置，能提供电泳时所需直流电。另有能调节电压和电流的输出装置。对常规电泳来说，输出电压在 100～500 V，电流在 150 mA 以内即合要求。

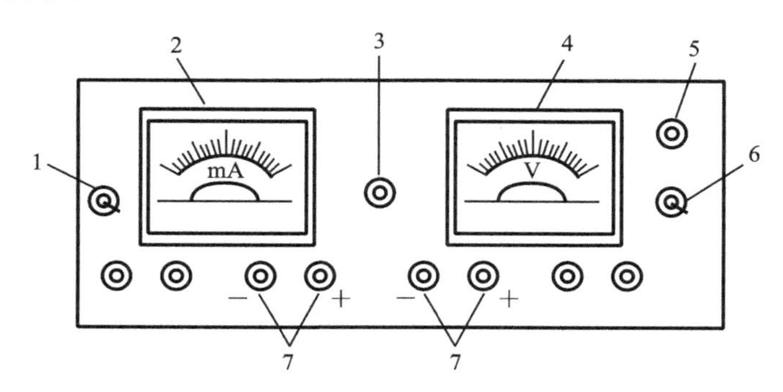

图 2-22　DY-1 型电泳仪的控制面板示意图

1—电流表量程转换开关(有“×1”和“×2”两挡)；2—直流电流表(0～200 mA)；3—输出电压调节器；
4—直流电压表(0～300 V)；5—指示灯；6—电源开关；7—输出插口(共 4 对)

DY-1 型卧式水平电泳槽适用于各种纸电泳、醋酸纤维薄膜电泳或载玻片电泳等，如图 2-23所示。电泳槽是用透明塑料模压(或用有机玻璃胶合)而成，中间有隔板，隔离成两个盛缓冲溶液的小槽，内装有电极(直径为 0.5 mm 左右的铂丝或镍铬钴合金丝)。槽口上方装有支撑条，支持物放在两个电极的两个支撑条上。支持物的两端分别浸入两个小槽的缓冲溶液中，形成盐桥。通电后，电流只能在支持物上通过，电泳物质即在其上泳动。

电泳仪的使用介绍如下。

1. 准备工作

(1) 检查电源电压，应为 220 V。

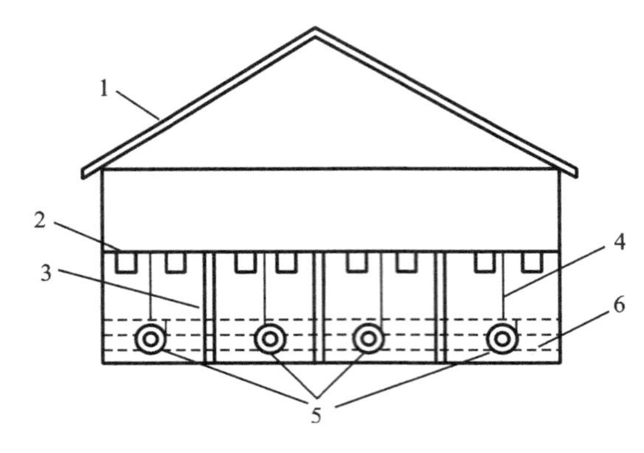

图 2-23　电泳槽示意图

1—电泳槽盖;2—支撑条;3—隔板;4—滤纸;5—电极;6—缓冲溶液

(2) 接线:将电源线的插头插入仪器背后的插座内,然后将一对输出线(视同时使用的电泳槽数而定)的一端插入仪器的输出插口内,另一端分别与电泳槽的正、负极相连。

(3) 在电泳槽内加入适量的缓冲溶液,并使两槽缓冲溶液在同一水平面上,放置好支撑条,中间隔板与两极支撑条的距离应相等。

(4) 在插上电源插头之前,应将面板上的开关关闭,将输出电压调节器按逆时针方向旋至最小,电流表量程开关拨至"×1"挡。

2. 操作方法及注意事项

(1) 开启电源开关,见指示灯亮。然后将输出电压调节器按顺时针方向慢慢旋转,即可见电压表和电流表的读数缓缓上升,一直调到所需的电压(或电流)值。电泳即开始进行。到预定时间后,关闭电源,停止电泳。

(2) 当输出的电流很大(如同时向几个电泳槽供电),超过 200 mA 时,可将电流表量程转换开关拨至"×2"挡,这时实际电流值应为电流表指示值的两倍。此数为几个电泳槽的电流值之和。

(3) 通电后,不要接触电泳槽内的电极及带电部分,以免触电。如需取放槽内物品或需插入或拔出输出插头,应将电源开关切断后进行,以免短路。

(4) 使用完毕,将输出电压调节器旋至最小,切断电源开关,拔去电源插头,做好仪器的清洁工作。

氨基酸在其等电点(pI)时呈两性离子状态,净电荷为零,在电场中既不向正极也不向负极移动。当溶液 pH 值小于 pI 时,氨基酸带正电荷,向负极移动;当溶液 pH 值大于 pI 时,氨基酸带负电荷,向正极移动。因混合氨基酸中各物质所带电荷性质、数量及相对分子质量不同,在同一电场中的泳动方向和速率不同,一定时间内各自移动的距离不同而得到分离。

三、仪器与试剂

【仪器】DY-1 型电泳仪,层析滤纸条,点样毛细管,镊子,铅笔,电吹风。

【试剂】0.2%天冬氨酸水溶液,0.2%精氨酸水溶液,天冬氨酸和精氨酸的混合液,茚三酮(按 0.2%的浓度溶解于缓冲溶液中),pH=8.6 的巴比妥缓冲溶液[1]。

【物理常数】

化 合 物	相对分子质量	熔点/℃	溶解性(H_2O)	等电点(pI)
天冬氨酸(Asp)	133.10	>300	溶	2.77
精氨酸(Arg)	174.20	244	溶(15%,21 ℃)	10.76

四、实验步骤

1. 点样和润湿

根据电泳槽的规格用镊子取三张适当长度和宽度的滤纸条[2],用铅笔在滤纸条中央轻轻画一横线,并在此横线上等距离处标上 Asp、AA、Arg 三个点,表示点样位置。用三支毛细管分别蘸取样品溶液,在上述位置分别点上天冬氨酸、混合氨基酸、精氨酸,样点的直径不超过2 mm。点样完毕,将滤纸放在电泳槽的支撑条上,两端浸入缓冲溶液中。当滤纸润湿至距离样品点 1 cm 左右时,用两把镊子按水平方向取出拉直,使缓冲溶液向中间样品扩散,直到两侧缓冲溶液同时接触到样品,待样品润湿后,将其夹在干滤纸中,吸去多余的缓冲溶液。注意整个过程不得用手接触滤纸条,以免污染滤纸。

2. 电泳

将润湿过的滤纸放在电泳槽的支撑条上,样品原点应在隔板正上方,滤纸两端浸入缓冲溶液中。如果一个电泳槽需放置多张滤纸,则滤纸之间不可接触,需有适当间隔。待全部滤纸条放妥后盖上槽盖,再进行通电操作,以免操作过程中触电。按电泳仪的使用方法连接导线进行电泳,慢慢调节输出电压调节器使输出电压稳定在 180 V,电泳约 60 min 后,关闭电源开关,拔出电源。

3. 显色

在滤纸条两端的两个电极液面处明显做出标记,以便量取支持物的有效长度。用两把镊子水平取出滤纸,置于洁净平面上[3],热风缓缓吹干至显色。记录显色斑点与原点的距离及电压和电泳的方向,计算各氨基酸的离子迁移率 μ,判断混合氨基酸的组成。

五、注解和实验指导

【注解】

[1] pH=8.6 的巴比妥缓冲溶液的配制:取二乙基巴比妥酸 1.84 g,二乙基巴比妥酸钠10.30 g,加蒸馏水至 1000 mL。最后加入 0.25 g 茚三酮用作氨基酸的显色剂。

[2] 滤纸条的长度需根据电泳槽的大小确定,以滤纸条两端能浸泡到两极缓冲溶液中 1~2 cm 为宜。滤纸宽度需根据点样数量的多少而定。若一张滤纸点一个样,以 1.5 cm 左右为宜;若一张滤纸点多个样,则需适当加宽,点多个样时以每个样点以及滤纸边缘之间距离为1 cm左右为宜。

[3] 不可随意放置于实验桌面,可用一张干净的纸垫着,以防滤纸污染后无法观察斑点。

【预习要求】

(1) 弄清氨基酸的等电点的含义,理解电泳法分离氨基酸的工作原理。

(2) 备好直尺和铅笔,设计好记录数据的表格。

【操作注意事项】

(1) 点样时不要用手触碰到纸条,否则皮肤上代谢的氨基酸会污染纸条。同时也不要接触到缓冲溶液,否则缓冲溶液中的茚三酮会使手上皮肤干燥后染色。

(2) 千万不要在通电状态下操作纸条,注意安全和各小组之间的配合。

六、思考题

(1) 什么是氨基酸的等电点? 等电点与氨基酸分子的带电状态有什么关系?

(2) 已知赖氨酸的 pI=9.74,苯丙氨酸的 pI=5.48,将它们置于 pH=8.9 的缓冲溶液中,它们各带什么电荷? 电泳时各向哪极移动?

Experiment 12　Paper Electrophoresis of Amino acids

Ⅰ Objectives

(1) To understand the principle of electrophoresis.

(2) To learn the paper electrophoresis method used for separation and identification of amino acids.

Ⅱ Principle

Electrophoresis is the study of the movement of charged species in an electric field. The technique, which has been used by biochemists for at least 60 years, is especially applicable to characterization and analysis of biological polymers.

1. Principle of electrophoresis

Molecular migration in an electric field is influenced by the size, shape, charge, and chemical nature of the charged species. The movement of a charged species (molecular or ion) subjected to an electric field is represented by Equation (2-6).

$$\mu = \frac{v}{E} = \frac{d/t}{V/l} = \frac{dl}{Vt} \tag{2-6}$$

Where, μ is the migration velocity($cm^2 \cdot V^{-1} \cdot s^{-1}$), V is the voltages on support(V), E is the strength of electric field($V \cdot cm^{-1}$), d is the migration distance of the molecule or ion (cm), v is the migration speed of the molecule or ion ($cm \cdot s^{-1}$), l is the effective distance of support(cm), t is the electrophoresis time(s or min).

2. Classification of electrophoresis

Electrophoresis techniques can be classified as either moving boundary electrophoresis (solution electrophoresis) or zone electrophoresis. According to the differences of sample supporters, the zone electrophoresis can be grouped into: paper electrophoresis; thin layer electrophoresis; cellulose acetate electrophoresis; starch gel electrophoresis; polyacrylamide gel electrophoresis (disc electrophoresis, planar plate electrophoresis, SDS-disc electrophoresis); agar electrophoresis.

3. Instrument and operation for electrophoresis

1) Electrophoresis instrument

The important features a satisfactory apparatus should possess are both constant voltage and amperage controls(both but cannot be operated simultaneously), which means that the apparatus include both a voltmeter and an ammeter covering the desired range. It has been stressed that the important figures are the voltage and amperage across the strip or gel. However, once the experiment has been standardized, the figures shown on the meters can be used to adjust the condition for subsequent runs. For general electrophoresis, the power pack has a range of 100～500 V at constant voltage with a maximum of 150 mA. A commonly used instrument for electrophoresis is shown in Figure 2-22.

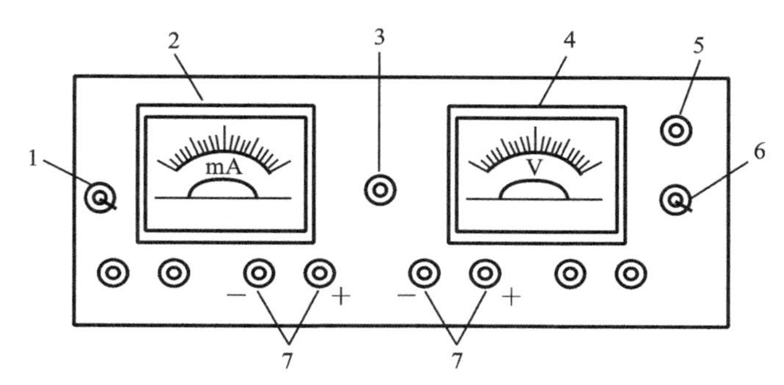

Figure 2-22　DY-1 Electrophoresis Apparatus Control Panel

1—Ammeter range switch(two ranges: "×1" and "×2"); 2—DC ammeter(0～200 mA); 3—voltage regulator; 4—DC voltmeter(0～300 V); 5—indicator light; 6—power switch; 7—output socket(4 pairs)

2) Electrophoresis tank

As shown in Figure 2-23, the tank base is a single moulded unit possessing two pairs of electrode-buffer compartments; the outer compartment of each pair contains the electrode wire. The paper strip is placed on the strip holders. When electrophoresis is running, the apparatus is covered with a transparent Perspex cover. At the end of the experiment the paper strip is transferred directly onto a clear filter paper for drying.

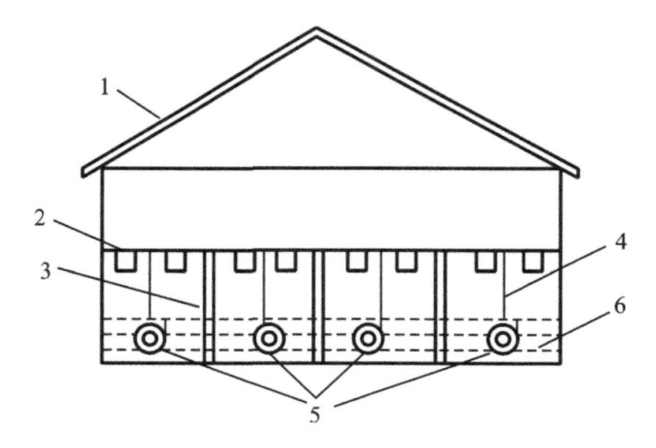

Figure 2-23　Schematic Diagram of Electrophoresis Tank

1—tank cap; 2—batten supporter; 3—clapboard; 4—paper strip; 5—electrode; 6—buffer solution

3) Operation on electrophoresis instrument

(1) Power pack is constructed to take input voltages of 220 V. And this should be checked before connecting a new apparatus to the mains.

(2) Connect the plug of electrophoresis instrument to the socket of the power, and insert the electrode to the instrument and another end of the electrode connected with the tank.

(3) Before turning on the power switch, rotate the voltmeter counterclockwisey to a minimum and the ammeter to "×1".

(4) Add the buffer to each compartment of the tank, until it reaches the height of the groove on the electrode.

(5) When wetting paper is ready, turn on the power switch, and then adjust control knob of voltmeter to the position where is proper for electrophoresis.

(6) Rotate the control knob of voltmeter to a minimum, and turn off the power after finishing electrophoresis.

4. Paper electrophoresis of amino acids

Amino acids contain both a proton acceptor ($—COO^-$) and a proton donor ($—NH_3^+$) group and exist as zwitterions (no net charge, no migration in electrophoresis) in the solid crystalline state. In solutions, however, an amino acid has several charged forms that are pH-dependant. At very low pH (plenty of free protons are available to neutralize $—COO^-$ group), amino acids exist as ions with an overall positive charge, while at high pH (plenty of free protons are available to neutralize $—NH_3^+$ group) they exist as ions with an overall negative charge. And thus, the charged ions are able to migrate to their charge opposite electrode in electrophoresis. The characteristic pH, at which the net electric charge is zero is called the isoelectric point (pI), can be used to estimate the charge on that amino acid at any pH in solution. If pH > pI, amino acids are negatively charged; if pH < pI, then positively charged.

At a curtain pH, the amino acids with different pI values in a mixture can take different charges in different quantity and kind, which lead to a detectable separation in electrophoresis.

Ⅲ Apparatus、Reagents and Materials

【Apparatus and materials】 Electrophoresis instrument, blower, capillary (diameter 1 mm), pencil, tweezer, ruler, filter paper (1.5 cm ×18 cm).

【Reagents】 0.2% aspartic acid (Asp) in H_2O, 0.2% Arginine (Arg) in H_2O, mixed solution of Asp and Arg, barbituric acid-barbiturate buffer solution (pH=8.6)[1], ninhydrin.

【Physical constants】

Compound	M_w	m. p. /℃	Solubility(H_2O)	pI
Aspartic acid	133.10	>300	Soluble	2.77
Arginine	174.20	244	Soluble(15%, 21 ℃)	10.76

Ⅳ Procedures

1. Spotting sample

Fold three pieces of filter paper with a tweezer and mark the fold with a light pencil line respectively[2]. The origin is thus at the mid-point of the paper strip. The spotting can be performed by the same techniques as for TLC by using capillary. One strip is for one sample. Each group needs 3 pieces of paper strips.

2. Wetting the paper

Place the spotted paper strips in the apparatus. The strips are tensioned across Perspex bridges so that they are fully stretched and do not sag in the middle. Two ends of each paper strip are dipped into buffer solution. Then the solvent rises up to the origin from two ends by capillary action.

3. Running electrophoresis

When the solvents in two ends meet at the origin, turn on the power switch and adjust the voltage to 180 V, running electrophoresis for 1 h.

4. Ceasing electrophoresis

Turn off the power switch, remove the paper strips with tweezers carefully out of the tank and transfer them directly onto a large piece of clean paper for drying by hot air with the blower until spots of the samples appear.

5. Recording data and calculating migration velocity

Measure the distance from origin to spots of the sample travelling, and record the voltage and time during electrophoresis. The orientation of sample travelling should be recorded as well. Calculate the migration velocity of each amino acid, and identify the composition of mixed amino acids.

Ⅴ Notes and Instructions

【Notes】

[1] Preparation of the buffer solution(pH=8.6): 1.84 g of two hexyl barbituric acid and 10.30 g of sodium two hexyl barbiturate are dissolved in 1000 mL of distilled water, then 0.25 g of ninhydrin is added.

[2] The length of paper strip is determined by the distance between the surfaces of buffer solution at two separated tanks. The strips should be able to touch with the solution at both ends.

【Requirements for preview】

(1) To understand how the different amino acids are separated by electrophoresis method.

(2) Prepare pencil and ruler and design a table for recording the lab data before the experiment.

【Experimental precautions】

(1) Don't touch the paper strips especially around the mid-part when you take them and spot the sample. Otherwise, the strip will be polluted by the amino acids on your hands.

(2) Never operate the electrophoresis instrument under the condition of "power on". Pay attention to safety and team cooperation.

Ⅵ Post-lab Questions

(1) What is the pI value of amino acids? What's the correlation between the pI value and the charges they take?

(2) When lysine(pI=9. 74) or phenylalanine(pI=5. 48) dissolve in the buffer solution with pH 8. 9, what charges do they have, respectively? What electrode do they migrate to?

Ⅶ Verbs

electrophoresis 电泳;
migration 迁移;
migration velocity 迁移速率;
strip 条,带;
compartment 分隔间;
aspartic acid 天冬氨酸;
arginine 精氨酸;
ninhydrin(水合) 茚三酮;
capillary 毛细管,毛细作用的;
barbituric acid 巴比妥酸;
barbiturate 巴比妥酸盐

第三部分　有机化合物的基本鉴定
——物理常数与性质鉴定

Part 3　Identification of Organic Compounds by Determination of Physical Constants and Characteristic Reactions

实验十三　固体有机化合物熔点的测定

一、实验目的

（1）了解测定熔点的原理和意义。
（2）掌握用毛细管法测定熔点的操作。

二、实验原理

结晶物质的熔点（T_m）是指在一个大气压下，固体物质与其熔融态达到平衡时的温度。在实验室通常采用毛细管法使用提勒管（Thiele tube）或者直接使用显微熔点仪进行测定，所需样品少，操作方便。实验室测定的熔点通常是指熔程，即结晶物质从开始熔化时到全部熔化时的温度范围。纯固体有机化合物一般有固定的熔点，结晶化合物的熔程一般不超过1 ℃，而含有杂质的固体化合物通常初熔点变低、熔程增长。因此，测定固体有机化合物的熔点不仅可以初步判断物质类别，也是确定化合物纯度的方便、有效的方法。

1. 毛细管法

毛细管法测定熔点装置如图3-1所示。首先把样品装入熔点管中。熔点管是一端开口、另一端封口的合适孔径的毛细管。将干燥的粉末状试样在表面皿上堆成小堆，将熔点管的开口端插入试样中，装取少量待测试样品的粉末。然后把熔点管竖立起来，开口端朝上，将一较大口径的玻璃管（长约70 cm）垂直于桌面放置，让熔点管在玻璃管内自由落下，这样重复几次，使样品装入熔点管底，样品高度为2～3 mm即可（见图3-1(a)）。为使测定结果准确，样品要研得极细，填充要均匀紧密。

将热浴液注入提勒管中，注入量为刚好淹没提勒管的上支管。热浴液可根据所测量物质的熔点选择。一般用液状石蜡（沸点220 ℃）、硅油（沸点250 ℃）等。毛细管中的样品应位于温度计水银球的中部（见图3-1(b)），可用乳胶圈捆好贴实，胶圈不要浸入热浴液中，以免软化脱落使热浴液变色。用有缺口的塞子套住温度计放到提勒管中，使温度计的水银球处在提勒管的两叉口之间，并让温度计刻度对着木塞缺口，以便读数（见图3-1(c)）。

在图3-1(c)中所示位置加热。载热体被加热后在管内呈对流循环，使温度变化比较均匀。在测定已知熔点的样品时，可先以较快速度加热，在距离熔点15～20 ℃时，应控制加热强度使温度计指数以每分钟 1～2 ℃的速度上升,直到熔化,测出熔程。加热速度过快会使熔程较大而得不到准确的结果[1]。在测定未知熔点样品时,应先粗测熔点范围,再按前述方法精测。测定时,应观察和记录样品开始塌落并有液相产生时(初熔)和固体完全消失时(全熔)的温度读数,所得数据即为该物质的熔程。还要观察和记录在加热过程中是否有萎缩、变色、发泡、升华及炭化等现象,以供分析参考。熔点测定至少要有两次重复数据,否则需再次测定。每次测定要用新毛细管重新装入新样品测定,不可将前面测定后留下的凝固物再次测量[2]。

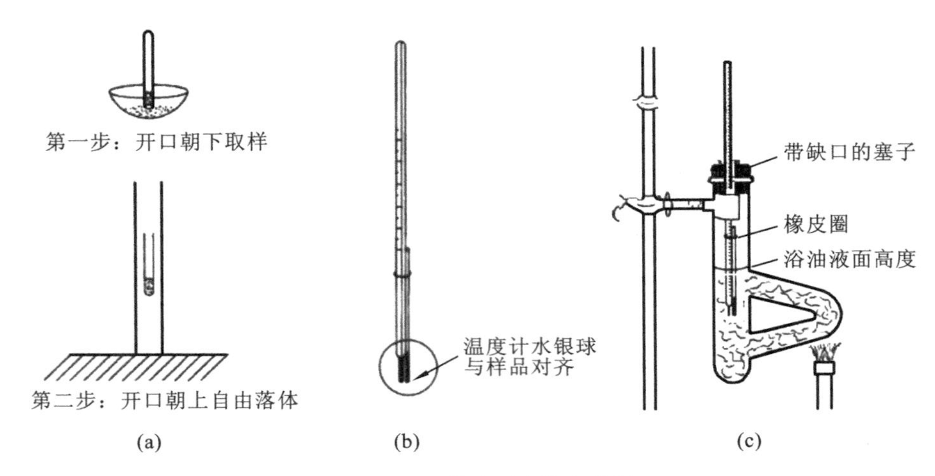

图 3-1　毛细管法测定熔点装置

2. 显微熔点仪测定熔点

这类仪器型号较多,因有半自动和全自动的不同而使操作有所差异,但共同特点是使用样品量少(2～3 颗小结晶),可通过显微镜直接观察晶体在加热过程中变化的全过程,如结晶的失水、多晶的变化及分解,能测量室温至 300 ℃样品的熔点。在使用这种仪器前必须仔细阅读仪器使用指南,严格按操作规程进行。

3. 温度计校正

测定熔点时,温度计上显示的熔点与真实熔点之间常有一定的偏差,为了测得精确的熔点,常常需要先对温度计进行校正。校正包括:在温度计的量程范围内,测量一系列标准物质的熔点,以测定值与标准值之间的差值为横坐标,以温度为纵坐标,从零到量程温度作图(温度计 0 ℃是精确的),以图中任一温度下的横坐标来消除用此温度计实际测量时的偏差。

三、仪器与试剂

【仪器】提勒管(b 形管),温度计(200 ℃),带缺口的橡皮塞,熔点毛细管,长玻璃管,玻璃棒,表面皿(或玻璃片),小胶圈,酒精灯,铁架台,铁夹。

【试剂】液状石蜡,苯甲酸苯酯,萘,乙酰苯胺,苯甲酸,尿素,水杨酸,对苯二酚。

【物理常数】

样品	苯甲酸苯酯	萘	乙酰苯胺	苯甲酸	尿素	水杨酸	对苯二酚
熔点/℃	68～70	80.5	114.3	122.1	131～135	157～159	172～175
闪点/℃	—	78.89	173.9	—	72.7	157	165
沸点/℃	298～299	217.9	304	249	196.6	211	285

四、实验步骤

（1）从提供的已知熔点的化合物中，选一到两种化合物，用提勒管装置测定熔点。粗测1次，精测2次，记录实测的物质熔程，并与标准值对比，分析对比结果。

（2）从已测过熔点的化合物中，任选一种化合物，掺杂5%～10%的另一种物质。混合均匀，测定样品熔点，验证杂质对熔点的影响。

（3）精确测定由指导教师提供的未知样品的熔点。这次测定，可以作为熔点测定技术的小测验。

（4）若实验条件允许，用提勒管装置和熔点仪分别测定相同样品的熔点，对比结果。

五、注解和实验指导

【注解】

[1] 晶体大小、晶形、加热速度和样品纯度都会影响熔点测定的准确性。热能从浴液传到样品，再到样品内部传导，都有时滞效应。因而晶体大小、晶形、加热速度会导致观察值与实际值不同。若加热太快，温度计读数将滞后于浴液温度导致测量值偏高。

[2] 样品冷却后的凝固可能因为晶形的改变而发生熔点的改变，故每次平行测定都需使用新样品。

【预习要求】

（1）了解熔点测定的意义及熔点测定中产生误差的各种原因与解决方法。

（2）了解熔程的概念，预习熔点测定的方法、原理。

（3）了解熔点测定装置及在操作中须注意的一些问题。

【操作注意事项】

（1）当使用煤气灯或酒精灯时，请将长发系于脑后，远离火焰；煤气灯或酒精灯不用时应及时关掉。

（2）加热含水的石蜡油或硅油时很不安全。含水的油加热到100 ℃（水沸点）时，由于形成水蒸气，热油将飞溅出来，油遇明火会被点燃。实验前不可冲洗空的提勒管，若不慎冲洗或检查提勒管底部已经有水层，则需更换提勒管，并将有水的提勒管交给指导教师。

（3）装样时均匀、紧密地装入2～3 mm高的样品。研磨和装填样品要迅速，防止样品吸潮。

（4）熔点测定操作的关键就是控制加热速度，使热量能均匀、平稳透过毛细管，保证样品受热熔化时，熔化温度与温度计所示温度一致。接近熔点时改用小火加热或将酒精灯稍微离开提勒管一些，尽量控制温度使之缓缓而均匀地上升。

（5）更换样品管时要等浴液冷却至低于熔点10～15 ℃，防止烫伤。

（6）实验结束后，浴液要冷却后方可倒回瓶中，提勒管不可水洗，直接放回指定地点。温度计不能马上用冷水冲洗，否则易破裂。

六、思考题

（1）指出下列陈述是否正确，并说明原因：

① 杂质会使有机化合物熔点升高。

② 像纯化合物一样，共熔体有尖锐的熔点。

③ 如油浴加热的速度太快,测定的熔点将偏低。

④ 熔点管内样品越多越好。

⑤ 用液状石蜡做油浴,不能测定熔点在 200 ℃以上固体的熔点。

(2) 在毛细管法测定熔点的装置中,为什么要用切口的塞子?

(3) 如何用混合熔点法判断两个固体样品是否可能为同一化合物?

Experiment 13 Determining Melting Point of Organic Solids

Ⅰ Objectives

(1) To know the identification of organic solids and their purity by determining the melting points.

(2) To practice determining melting point of organic solids by micro melting-point method.

Ⅱ Principle

The melting point is the temperature at equilibrium when starting in the solid state and going to the liquid state under normal atmosphere. In all micro methods the melting point is actually determined as a melting range. Melting points of pure substances occur over a very narrow range and are usually quite sharp. The criteria for purity of a solid are the narrowness of the melting point range and the correspondence to the value found in the literature. Impurities will lower the melting point and cause a broadening of the range. For example, pure benzoic acid has a reported melting point of 122.1 ℃; benzoic acid with a melting point range of 121～122 ℃ is considered to be quite pure. The broadening of the melting range that results from introducing an impurity into a pure compound may be used for identifying if a substance is pure or not.

The micro melting-point determination, which uses a very small amount of sample, is convenient and easy to see. The sample is heated slowly in a special apparatus, which is "b"-shaped and called as Thieletube, equipped with a thermometer and heating oil as a heating bath [1]. The commonly used heating oil is liquid paraffin wax oil(b. p. 220 ℃)or silicone oil (b. p. 250 ℃). In determining, two temperatures are noted. The first is the point at which the first drop of liquid forms among the crystals, and the second is at which the whole mass of crystals turns to a clear liquid. The melting point is then recorded giving this range of melting. Note that some melting point data in handbooks are the average of the actual lab figures.

1. Capillary tubes and sample preparation

Capillary tubes used for melting-point determination have one end already sealed and are open at the other end to permit introduction of the sample. The sample is put into the tube as follows(see Figure 3-1(a)). Place a very small amount of the solid sample in a clean watch glass and tap the open end of the capillary tube into the solid in the glass so that a small amount is forced 2～3 mm into the tube, take a piece of 70～80 cm long glass tube, place this tube vertically on a hard surface, and drop the capillary tube (sealed end down) through the large tube several times (see Figure 3-1(a)). And then you might see that the solid end up packed at the sealed end of the capillary tube. Prepare 2～3 pieces of sample at one time for the parallel measurements [2].

2. Thiele tube apparatus

A simple type of melting-point apparatus is the Thiele tube. By heating the point at the side arm, the heating oil in this "b"-shaped tube can distribute the heat to all parts of the vessel and stirring is not required. Proper use of the Thiele tube is crucial to obtain reliable melting points. The general procedures are as follows.

(1) Add some liquid oil as a heating bath into a dry Thiele tube. Support the apparatus on a clamp. Make sure that the height of the heating oil is at the level indicated in Figure 3-1(c).

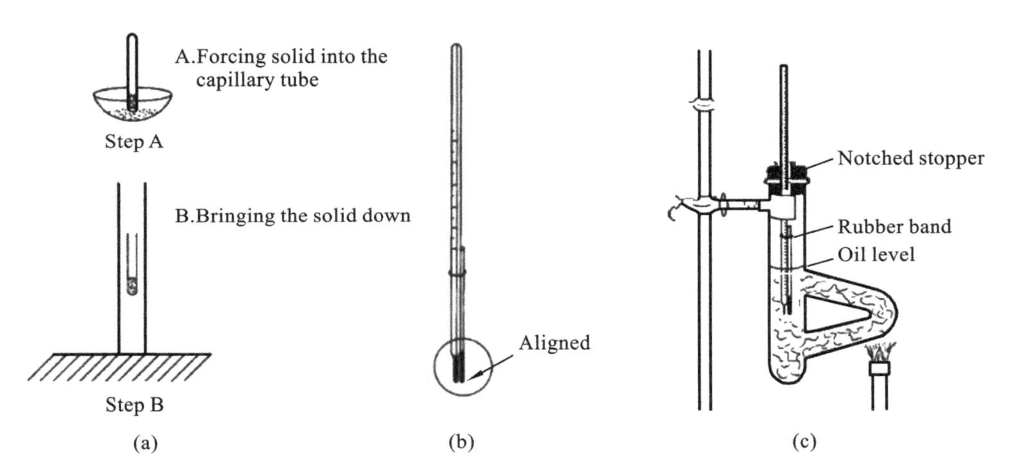

Figure 3-1　Sample Preparation and Thiele Tube Apparatus for Micro Melting-point Method

(2) Attach the capillary tube containing the sample to the thermometer by means of a rubber band. The sample itself should be directly adjacent to the bulb of the thermometer.

(3) Support the thermometer in the Thiele tube with a notched rubber stopper and make sure that the bulb of thermometer should be in correct position as indicated (see Figure 3-1 (b)). Note that the rubber band should remain above the level of heating fluid to avoid contacting with the hot oil. The stopper is cut away on one side so as to make visible the thermometer markings in that vicinity. This cut also serves the purpose of making the apparatus an open system. Never heat a closed system!

(4) Heat the Thiele tube by a burner. Apply a strong heating before the temperature is near to the approximate melting point (10～15 ℃ below melting point) and after that a gentle heating at the rate of 1～2 ℃ per minute is crucial to get an accurate melting point.

Ⅲ Apparatus and Reagents

【Apparatus】 Thiele tube, capillary tubes, thermometer (～200 ℃), glass tube (70～80 cm), rubber band, rubber stopper(notched).

【Reagents】 Paraffin wax oil, phenyl benzoate, naphthalene, acetylaniline, benzoic acid, urea, salicylic acid, 1,4-benzenediol.

【Physical constants】

Compound	Phenyl benzoate	Naphthalene	Acetyl-aniline	Benzoic acid	Urea	Salicylic acid	1,4-Benzenediol
m. p. /℃	68～70	80.5	114.3	122.1	131～135	157～159	172～175
Flash Point/℃	—	78.89	173.9	—	72.7	157	165
b. p. /℃	298～299	217.9	304	249	196.6	211	285

Ⅳ Experiments

(1) Select one or two compounds from a list of supplied compounds with known melting points. Determine the melting points (ranges) for each of these substances using the capillary tube and the Thiele tube apparatus. Repeat 2～3 times of measurements for each compound. Compare your results with known melting points.

(2) Select two samples measured from above and mix them thoroughly in 9 : 1 ratio. Determine the melting range of the mixed sample in order to verify the anticipated effects of impurities on the melting range of a pure substance.

(3) Accurately determine and report the melting range of an unknown sample supplied by your instructor. This determination may be considered as a practical lab quiz.

(4) If permitted, determine the same sample by Thiele tube apparatus and melting point instrument respectively for a comparison between the two methods.

Ⅴ Notes and Instructions

【Notes】

[1] The accuracy of determination is crucially influenced by heating control. Heat is distributed from the bath to the sample. There is time lag effect in the convection, which leads to the melting behind thermometer reading.

[2] The melting point is slightly changed with changed crystal size and morphology. For this reason, those heated sample tubes cannot be reused for a second measuring, because the crystal size and morphology would have changed when the melting sample crystallized during the cooling.

【Requirements for preview】

(1) To understand the significance of melting point determination and various influencing factors of determination accuracy and solution.

(2) To establish the concept of melting range and learn the principle of determination.

(3) To learn about some problems probably appear in the operation.

【Experimental precautions】

(1) Please confine long hair to keep away from the flame, the burner should be turned

off if not in use.

(2) Heating the water-containing bath oil is dangerous. Do not wash the Thiele tube by water and check if there is an aqueous layer at the bottom of Thiele tube before use. In case that the Thiele tube has been rinsed by accident, it will be very necessary to change a new one and hand in the old one to your instructor.

(3) Grind and pack samples quickly to prevent moisture absorption. The height of sample in the capillary tube is 2～3 mm.

(4) Temperature control is accomplished by adjusting the flame or taking the burner a little away from the heating position.

(5) Wait for the hot oil cooling down for a while (10～15 ℃ below melting point at least) before apply next measuring. Take care not being hurt when replacing the sample tube.

(6) When the experiments finished, pour back the bath oil after it cooling down and return the Thiele tube apparatus without washing.

Ⅵ Post-lab Questions

(1) Indicate which of the following statements is true(T)or false(F). Why?

① An impurity raises the melting point of an organic compound.

② A eutectic mixture has a sharp melting point, just as does a pure compound.

③ If the heating rate is too fast, the melting point that results will likely be too low.

④ The more the sample used in the capillary tube, the better to the micro melting-point determination.

⑤ A heating bath containing mineral oil should not be used to determine the melting points of solids melting above 200 ℃.

(2) In the melting-point determination by Thiele tube apparatus, why should the stopper be notched?

(3) How to find out if two unknown samples are the same compounds or not?

Ⅶ Verbs

micro melting-point method 微量熔点法；
Thiele tube 提勒管；
capillary tube 毛细管；
paraffin wax oil 石蜡油；
silicone oil 硅油；
rubber stopper 橡皮塞

实验十四　液体有机化合物折光率的测定

一、实验目的

(1) 了解测定折光率对研究有机化合物的意义。

(2) 掌握使用阿贝折光仪测定液体化合物折光率的方法。

二、实验原理

1. 基本原理

折光率(refractive index)是有机化合物的重要物理常数之一。固体、液体和气体都有折光率,尤其是液体有机化合物,文献记载更为普遍。通过测定折光率,可以判断有机化合物的纯度,也可以鉴定未知物。

在不同介质中,光的传播速率是不同的。当光从一种介质射入另一种介质时,若它的传播方向与两种介质的界面不垂直,则其在界面处的传播方向会发生改变。这种现象称为光的折射(见图 3-2)。

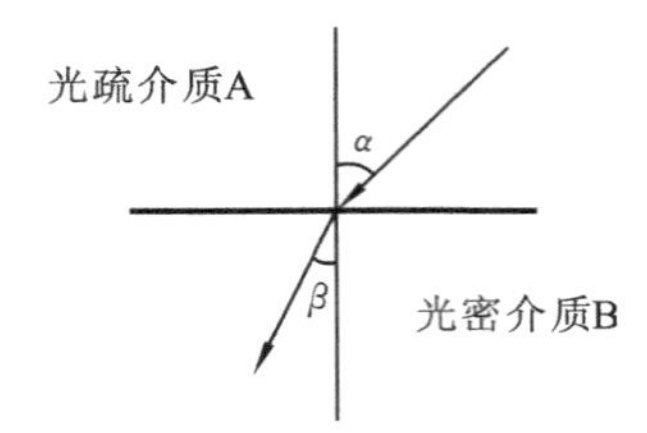

图 3-2　光的折射现象

根据折射定律,波长一定的单色光在确定的外界条件下(如温度、压力等),从一种介质 A 射入另一种介质 B 时,其入射角 α 的正弦与折射角 β 的正弦之比和两种介质的折光率成反比,即

$$\sin\alpha/\sin\beta = n_B/n_A \quad (3\text{-}1)$$

若设定介质 A 为光疏介质,介质 B 为光密介质,则当入射角 $\alpha=90°$,即 $\sin\alpha=1$ 时,折射角最大,称为临界角,以 β_0 表示。折光率的测定可近似地视为在真空状态中,即 $n_A=1$,则 $n_B=1/\sin\beta_0$。因此,通过测定临界角 β_0,即可得到介质 B 的折光率(n)。

折光率与物质结构、入射光线的波长、温度、压力等因素有关。通常大气压的变化影响不明显,只是在精密测定时才考虑。使用单色光要比用白光测得更为精准,因此常用钠光(D 线,$\lambda=589.3$ nm)做光源。折光率(n)的表示需要注明所用光线波长和测定的温度,常用 n_D^{20} 来表示,即以钠光为光源,20 ℃时所测定的 n 值。

通常温度升高(或降低)1 ℃时,液态有机化合物的折光率就减少(或增加)$3.5\times10^{-4}\sim5.5\times10^{-4}$(通常取均值 4.5×10^{-4}),在实际工作中常采用粗略的换算公式把某温度下所测得的折光率换算成另一温度下的折光率,换算公式为

$$n_D^T = n_D^t + 4.5\times10^{-4}(t-T) \quad (3\text{-}2)$$

式中,T 为规定温度(℃),t 为实际测量时的温度(℃)。

2. 阿贝折光仪

根据不同的需求,市面上的折射仪有手持式折光仪、糖量折光仪、蜂蜜折光仪、宝石折光仪、数显折光仪、全自动折光仪及在线折光仪等。折光仪主要由高折光率棱镜(铅玻璃或立方氧化锆)、棱镜反射镜、透镜、标尺(内标尺或外标尺)和目镜等组成。在有机化学实验室里,一般用阿贝折光仪(Abbe refractometer)来测定折光率(见图 3-3),其工作原理就是基于光的上述折射现象,通过目视望远镜部件和色散校正部件组成的观察部件来调节并找到明暗分界线视场,也即临界角的位置(见图 3-4),并由数据处理系统将角度转换成折光率显示出来。

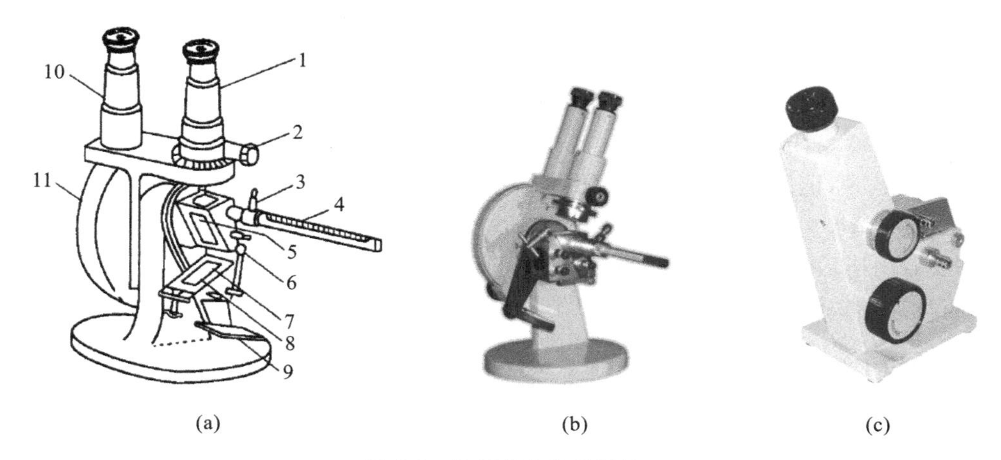

图 3-3　阿贝折光仪的结构

1—镜筒；2—光散射调解旋钮；3—恒温水主入口；4—温度计；5、6、7—棱镜组；8—样品槽；9—反射镜；10—折光率读数观察目镜；11—刻度盘

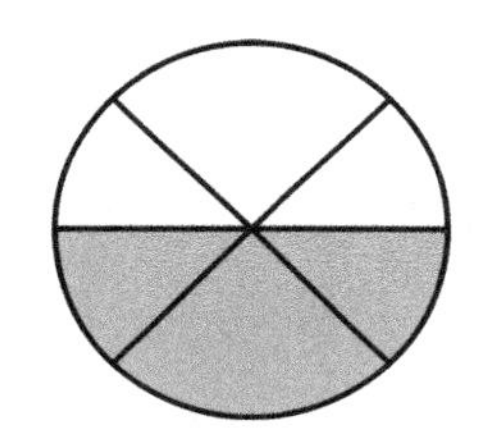

图 3-4　阿贝折光仪在临界角时目镜视野图

三、仪器、试剂与材料

【仪器】阿贝折光仪。

【试剂与材料】无水丙酮，双蒸水，无水乙醇，乙酸乙酯，松节油；擦镜纸。

【物理常数】

物质名称	水溶性	密度/(g·cm^{-3})	沸点/℃	折光率(n_D^{20})
丙酮	混溶	0.7920	56	1.3589
纯水	—	1.0000	100	1.3330
乙醇	混溶	0.7893	78.3	1.3614
乙酸乙酯	8.5(15 ℃)	0.9020	77.2	1.3723
松节油	0	0.85～0.87	154～165	1.4670～1.4710

四、实验步骤

(1) 开启恒温水浴使温度恒定至 20 ℃，若未连接恒温水浴，则此项操作免去。

(2) 仪器校正。当恒温后，松开锁钮，开启棱镜，滴入 1～2 滴丙酮[1]于镜面上，用擦镜纸蘸少许丙酮轻轻擦拭上、下两棱镜镜面。待完全干燥后，在折射棱镜的抛光面上滴 1～2 滴高纯度蒸馏水，盖上进光棱镜。调节反光镜，使镜筒视场最亮[2]。通过目镜观察视场，同时旋转调节手轮和色散校正手轮，使在目镜中观察到明、暗两部分具有良好的反差和明暗分界线具有最小的色散,视场内明暗分界线准确对准交叉线的交点(见图 3-4)。如有偏差,则可用钟表螺丝刀通过色散校正手轮中的小孔,小心旋转里面的螺钉,使分界线相位上下移动至交叉线的交点,然后再进行测量,直到读出的纯水的折光率符合标准为止。

(3) 打开棱镜,测定无水乙醇、乙酸乙酯和松节油的折光率。当视场内明暗分界线准确对准交叉线的交点时,记录从镜筒中读取的折光率,同时记下温度。重复测定 2～3 次,取其平均值为样品的折光率。

(4) 仪器用毕后用沾有少量丙酮的擦镜纸擦拭干净上、下镜面,晾干后合紧两面,用仪器罩盖好,将废弃的擦镜纸收进垃圾桶。

五、注解和实验指导

【注解】

[1] 丙酮具有良好的溶解性能和易挥发的特性,适用于镜面的清洁,也可用乙醇或乙醚代替。

[2] 较为简易的阿贝折光仪采用的是自然光源,需要通过反光镜将光线反射入棱镜,此时需要将折光仪放置在光线充足的水平台上。

【预习要求】

(1) 了解折光率的物理意义及阿贝折光仪的结构特点。

(2) 理解仪器校正与样品测量的异同点。

【操作注意事项】

(1) 仪器使用前后及更换样品时,必须用丙酮或乙醇清洗干净折射棱镜系统的工作表面并干燥,以防留有其他物质,影响成像清晰度和测量精度。

(2) 不可测定强酸、强碱等具有腐蚀性的液体。

(3) 保持仪器清洁,严禁油手或汗手触及光学零件。

(4) 阿贝折光仪最重要的部件是一对棱镜,使用时应注意保护棱镜,擦镜面时只能用擦镜纸而不可用滤纸等。加试样时切勿将管口触及镜面。滴管口要烧光滑,以免不小心碰到镜面造成刻痕。

(5) 读数时,有时在目镜中看不到半明半暗界线而是畸形的,这是由于棱镜间未曾充满液体;若出现弧形光环,则可能是有光线未经过棱镜而直接照射在聚光透镜上。

六、思考题

(1) 液体化合物的折光率与什么因素相关?

(2) 已知某化合物在 20 ℃时的折光率为 1.3578,请预测在室温为 26 ℃时测定的折光率有何变化。

Experiment 14　Determining Refractive Index of Organic Liquids

Ⅰ Objectives

(1) To learn the significance of refractive index in research of organic liquids.

(2) To master the determination of refractive index by using an Abbe refractometer.

Ⅱ Principle

1. Definition of refractive index

Refractive index is an important physical constant of organic compounds. The compounds in all possible states—gas, liquid or solid have their specific refractive index. The determination of refrative index is especially useful and simple for identifying liquids or indicating their purity.

Refractive index(n), also called index of refraction, a number indicating the speed of light in a given medium, usually is represented as the ratio of the speed of light in a vacuum or in air to that in the given medium(ex. organic liquids)(see Figure 3-2). It is a measure of the extent to which radiation is refracted on passing through the interface between two media. It can be described by the ratio of the sine of the angle of incidence(α) to the sine of the angle of refraction(β), which can be shown to be equal to the ratio of the phase speed in the first medium to that in the second. This is described by Equation(3-1).

$$\sin\alpha/\sin\beta = n_B/n_A \qquad (3\text{-}1)$$

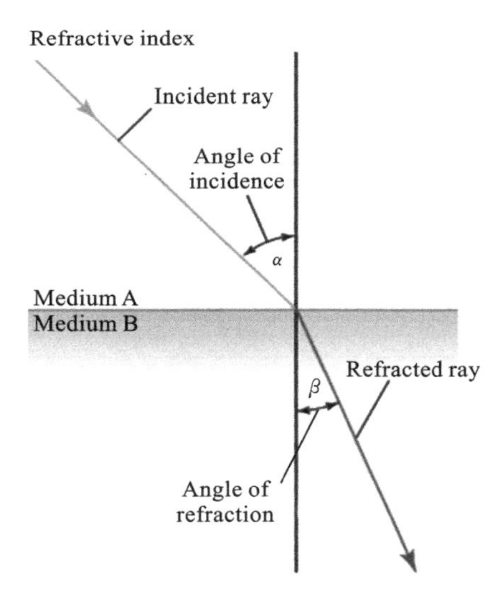

Figure 3-2 The Refraction of Light when Passing through the Interface between Two Media

If $\alpha = 90°$, that is, $\sin\alpha = 1$, then the angle of refraction(β) is a maximum, designated as β_0. And in the air or in the vacuum, $n_A \approx 1$. So, $n_B = 1/\sin\beta_0$. That means, we can get n_B by measuring the maximum angle of refraction, β_0.

Refractive index of materials varies with the wavelength of the beam of light and temperature. It is usual to report refractive index measured at 20 ℃, with a sodium lamp as the source of illumination. The sodium lamp gives off yellow light of 589.3 nm at wave length, the so-called sodium D line. Under these conditions, the refractive index is reported in the form of n_D^{20}.

Refractive index of liquid always decreases as temperature increased. Variations due to change in temperature are somewhat dependent on the class of compound observed, but are usually somewhere between 3.5×10^{-4} and 5.5×10^{-4} per ℃. Taking the average value of 4.5×10^{-4} serves as a fair approximation for most liquids, the corrected refractive index is given as below:

$$n_D^T = n_D^t + 4.5 \times 10^{-4}(t - 20) \quad (3\text{-}2)$$

Where T is the given temperature(℃), t is the thermometer temperature(℃).

2. The Abbe refractometer

The instrument used to measure the refractive index is called refractometer. There are many types of refractometers on the market under different demands. The most common instrument is the Abbe refractometer with one or two eyepieces and digital display. The Abbe refractometer mainly consists of a highly refractive prism, a reflecting prism, an optical lens, an inner or outer scale plate and eyepieces. A common type of Abbe refractometer in organic lab is shown in Figure 3-3. Its general working principle is based on the above mentioned phenomenon of refraction of light when passing through from air to the liquid. Refractive index can be read conveniently from the eyepieces after adjusted correctly.

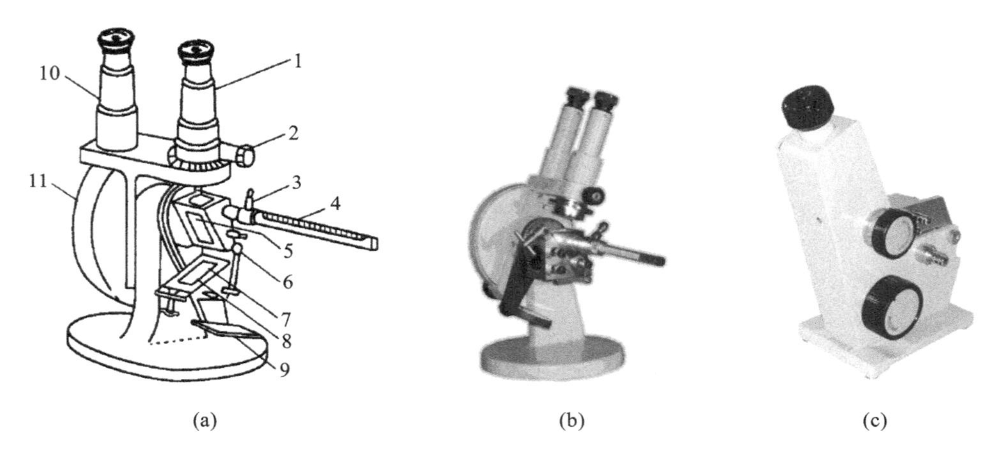

Figure 3-3 Abbe Refractometer

1—eyepiece with lens; 2—chromatic adjustment collar; 3—water; 4—thermometer; 5, 6, 7—hinged prism; 8—groove for sample; 9—reflector; 10—lens to view index of refraction scale; 11—index of refraction scale

Ⅲ Apparatus, Reagents and Materials

【Apparatus】 Abbe refractometer.

【Reagents and materials】 Anhydrous acetone, distilled water, anhydrous ethanol, ethyl acetate(EA), turpentine oil; lens paper.

【Physical constants】

Compound	Water Solubility/[g · (100 mL)$^{-1}$]	ρ/(g · cm^{-3})	b. p. /℃	n_D^{20}
CH_3COCH_3	Miscible	0.7920	56	1.3589
H_2O	—	1.0000	100	1.3330
CH_3CH_2OH	Miscible	0.7893	78.3	1.3614
$CH_3COOC_2H_5$	8.5(15 ℃)	0.9020	77.2	1.3723
Turpentine oil	0	0.85～0.87	154～165	1.4670～1.4710

Ⅳ Procedures

(1) Begin circulation of water from the thermostatic water bath well to get a constant temperature in advance. (If not connected, please skip this step.)

(2) Instrument calibration.

① Swing open the upper prism gently and check the surface of the prisms for cleaning [1]. Place 1～2 drops of acetone on the surface of prism and wipe the upper and lower surfaces with lens paper. When the surfaces are dry, drop 2～3 drops of distilled water onto the lower prism. Lower the upper prism and lock it into position.

② Look into the eyepieces. Rotate the reflector so it shines through the prism into the sample area.

③ Now adjust the light and the chromic adjustment knob until the field seen in the eyepieces is illuminated so that the light and dark regions are separated by as sharp a boundary as possible (see Figure 3-4). If the boundary has colors associated with it and/or appears somewhat diffuse, rotate the compensator drum on the face of the instrument until the boundary becomes noncolored and sharp.

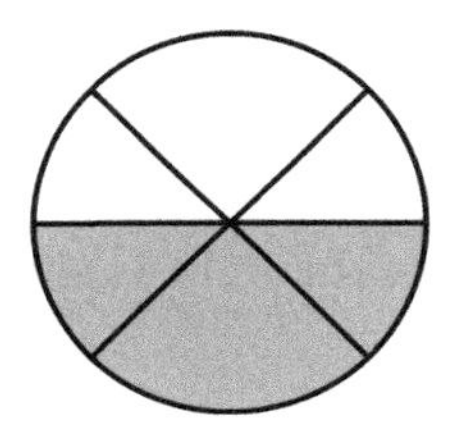

Figure 3-4　View through the Eyepieces when the Refractometer Adjusted Correctly

④ Read the value that appears on the field. Reported n_D^{20} of distilled water is 1.3330. If there exists an error for the reading, then slowly rotate the screw of the chromic adjust knob from a hole inside until the reading is calibrated to the standard.

(3) Now repeat above procedures. Measure the refractive index of ethanol, ethyl acetate (EA) and turpentine oil respectively. Take several replicate readings and report the average value.

(4) Taking care not to scratch the surfaces, clean the refractometer prism faces with a soft lens paper moistened with acetone immediately after use. Cover the refractometer before leave.

Ⅴ Notes and Instructions

【Notes】

[1] Acetone, as a good volatile organic solvent, is very suitable for cleaning the surface of prisms. The other similar solvents such as ethanol and diethyl ether are also often used for cleaning.

[2] Relatively simple Abbe refractometer adopts room light as the source of illumination, which needs to rotate the reflector to induce the light into the prism. So it's

better to place the refractometer on a horizontal table with exposure to sufficient light.

【Requirements for preview】

(1) Get to know the physical meaning of n_D^{20} and the working principle of the Abbe refractometer.

(2) To understand the differences and similarities between the two steps of instrument calibration and sample measurement.

【Experimental precautions】

(1) The working surface of prisms must be cleaned with acetone or ethanol before and after use in case that the residues may affect the imaging resolution and precision of the instrument.

(2) Those strong acidic or basic liquids, which are highly corrosive and harmful to the prisms, should be far away from the refractometer.

(3) Avoid touching the optical system with dirty hands.

(4) The protection of the prisms is the most important. The lens paper must be choosed instead of filter paper when cleaning the surface of prism. Avoid touching the surface when place the samples by a glass dropper. Otherwise, the sharp edge of the glass might scratch and hurt the surface.

(5) If the boundary of the light and dark regions is abnormal and cannot be adjusted to a straight line, it is probably because the gap between two prisms is not filled with liquid. If curved aureole is viewed, that is because the light is shining directively on the lens not via the reflecting prism.

Ⅵ Post-lab Questions

(1) What are the influencing factors for the refractive index of liquids?

(2) How does the value of refractive index change at 26 ℃ for a compound($n_D^{20}=1.3578$)?

Ⅶ Verbs

refractive index 折光率;

Abbe refractometer 阿贝折光仪;

prism 棱镜;

reflecting prism 反射棱镜;

optical lens 光学透镜;

eyepiece 目镜

实验十五　旋光度的测定

一、实验目的

(1) 掌握使用旋光仪测定手性物质旋光度的方法。

(2) 学习比旋光度的计算方法。

二、实验原理

手性化合物使平面偏振光的振动平面向右(顺时针)或者向左(逆时针)旋转的性质叫做旋

光性，发生偏转的角度称为旋光度。影响旋光度的因素包括旋光物质的浓度、样品管的长度、溶剂、温度及光源波长。为了比较不同旋光物质的旋光性能，需要有一个统一的比较标准，这个标准就是比旋光度$[\alpha]_D^t$，它是指在一定测定温度、光源和溶剂条件下的单位旋光度。比旋光度是手性物质特有的物理常数之一，手册、文献上多有记载。测定比旋光度可以鉴定旋光物质的纯度和含量。

当被测物为纯液体时，$[\alpha]_D^t=\frac{\alpha}{l \cdot \rho}$，故液体的密度 $\rho=\frac{\alpha}{l \cdot [\alpha]_D^t}$。

当被测物为固体时，需配成溶液来测[1]，$[\alpha]_D^t=\frac{\alpha}{l \cdot C}\times 100\%$，则物质含量 $C=\frac{\alpha}{l \cdot [\alpha]_D^t}$。

其中，l 为旋光管的长度(dm)，ρ 为密度($g \cdot cm^{-3}$)，C 为溶液浓度($g \cdot (100\ mL)^{-1}$)。注意计算时将字母单位统一换算。

1. 旋光仪

测定旋光度的仪器称为旋光仪，它主要由光源、起偏镜、样品管、检偏镜、目镜和游标刻度盘等部件组成。一般实验室使用的是目测旋光仪，其仪器外形及基本构造如图 3-5 所示。

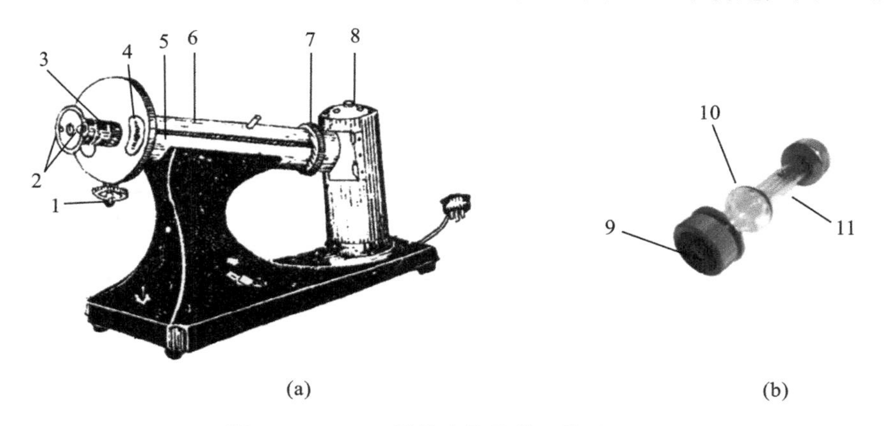

图 3-5　WXG-4 型旋光仪及样品管外形图

1—刻度盘转动手轮；2—放大镜座；3—焦距调节螺旋；4—游标刻度盘；5—起偏镜；
6—镜筒盖；7—检偏镜；8—钠灯；9—玻璃透光片；10—样品池凸起(用于聚集气泡)；11—样品池

测量时，可从目镜中看到一个明暗相间的三分视场，如图 3-6(a)或(b)所示。通过旋转刻度盘来调整，使视场的中间部分与其两边部分没有明显的界线，强度均匀且整个视场较暗(见图 3-6(c))，判断的标准是稍微来回旋转检偏镜时可观察到三分视场(a)和(b)来回转变，再把图像调回到中间无明显明暗界线的较暗的均一视场[2]，记下刻度盘读数 α。

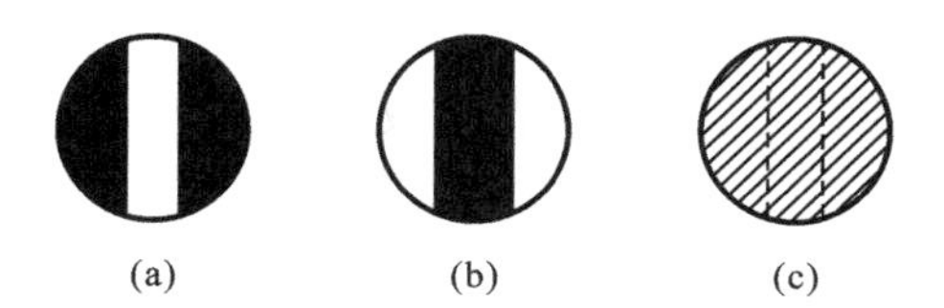

图 3-6　三分视场变化示意图

2. 读数方法

两边的游标刻度盘上分别标出 0°～180°，并有固定的游标，它分为 20 等分，等于刻度盘 19 等分。读数时先看游标的“0”落在刻度盘上的位置，记下整数值，再利用游标尺与主盘上刻度线重合的位置,读出游标尺上的数值(为小数),可以准确读到两位小数。为了消除刻度盘偏心差,可采用双游标读数法,并按下列公式求得结果:

$$Q=(A+B)/2$$

式中,A 和 B 分别为两游标窗读数。如果 $A=B$,而且刻度盘转到任意位置都符合等式,则说明仪器没有偏心差。读数示意图见图 3-7。

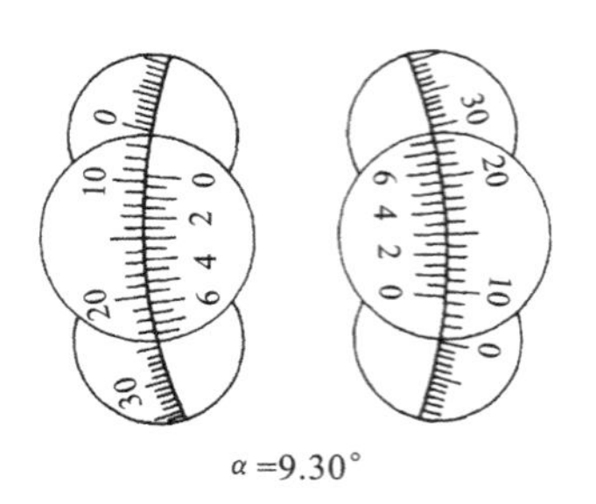

图 3-7　游标盘刻度读数示意图

三、仪器、试剂与材料

【仪器】旋光仪,样品管。

【试剂与材料】双蒸水,葡萄糖溶液,果糖溶液;擦镜纸。

【物理常数】

化　合　物	H_2O	D-(+)-葡萄糖	D-(－)-果糖
比旋光度$[\alpha]_D^{20}$	0	+52.25°	－ 92.25°

四、实验步骤

(1) 将仪器接至 220 V 交流电源上,打开电源开关,钠光灯亮,预热 5 min。

(2) 将三支样品管分别装入蒸馏水、葡萄糖溶液和果糖溶液。样品管中若有气泡,让气泡浮在样品管的突出处,用软布拭净样品管两端,使之有较好的透光性。样品管螺帽不宜旋得过紧。

(3) 打开镜盖,将盛有蒸馏水的样品管放入镜筒中,盖上镜盖。看着目镜内同时调节焦距旋钮,使镜筒中三分视场画面锐利清晰。旋转刻度盘手轮直至三分视场各部分明暗程度完全一致。准确读取并记录刻度盘读数,该值为零点校正读数,用于消除系统误差。

(4) 再分别将装有葡萄糖溶液和果糖溶液的样品管置于镜筒中同上测定,准确读取刻度盘读数,结合蒸馏水的零点校正读数,确定葡萄糖溶液和果糖溶液的旋光度。

(5) 根据所测旋光度值计算葡萄糖和果糖溶液的浓度。

五、注解和实验指导

【注解】

[1] 旋光度与光束通路中光学活性物质的分子数成正比。比旋光度值较小或溶液浓度小的样品,在配制待测样品溶液时,宜将浓度配高一些,并选用长一点的旋光管,以便观察。

[2] 将刻度盘旋转 180°可看到有两个使三分视场明暗一致的读数,一个视场较暗,一个视场很亮。应取较暗的视场读数,此读数灵敏、准确。而整个视场很亮的读数误差较大。

【预习要求】

（1）学习旋光度与比旋光度、手性与非手性、光学活性、对映异构等基本概念。

（2）了解旋光仪的构造和工作原理。

【操作注意事项】

（1）目测型旋光仪在测量前需要预热一段时间才能使光线亮度达到稳定要求。

（2）装样后的样品管从两端目测不能有大气泡，两端的玻片需擦拭干净，否则会造成光线的折射而产生测量误差。

（3）因需要人为找出均一视场，最好多平行测定几次求平均值，以消除视觉误差。

（4）蒸馏水的校正值需要在计算时扣除，若校正值为负值，则需要减去这个负值。

（5）须记录所用旋光管的长度、测定温度及所用溶剂（如用水做溶剂则可省略）。

六、思考题

（1）测量旋光度时，为何要用纯溶剂做空白对照进行零点校正？

（2）如何确定某溶液的旋光度为－10°而不是＋170°？

Experiment 15　Determination of Rotation Angle

Ⅰ Objectives

(1) To know the measurement of optical rotation of chiral compounds by using the polarimeter.

(2) To learn the meaning of specific rotation.

Ⅱ Principle

Chiral compounds distinguish themselves with other compounds by their ability of change the oscillating direction of a beam of plane-polarized light. They can rotate the oscillating plane to the right (labeled as "＋") or to the left ("－"), and the extent of the rotation is measured as rotation angle (α). The rotation angle for one compound is influenced by a series of factors: total sum of the molecule numbers; temperature; wavelength of the light source. All these factors should be taken into account in the measurement. In order to estimate the comparable potency of optical rotation between different chiral compounds, all the influencing factors should be placed at a standard level in measuring, and the rotation measured under such standard conditions is called specific rotation, labeled as $[\alpha]_D^t$. Specific rotation is a characteristic physical constant of chiral substances which is recorded by most of the handbooks. The determination of rotation angle can be used to identify the optical purity of chiral compounds or calculate their contents in the solution.

When the sample is liquid, $[\alpha]_D^t=\frac{\alpha}{l\cdot\rho}$, so the liquid density $\rho=\frac{\alpha}{l\cdot[\alpha]_D^t}$.

When the sample is solid, it should be dissolved in a certain solvent for measuring, $[\alpha]_D^t=\frac{\alpha}{l\cdot C}\times100\%$, then the sample content $C=\frac{\alpha}{l\cdot[\alpha]_D^t}$.

Where "l" is the length of sample tube(dm), ρ is the density of liquid sample($g \cdot cm^{-3}$), C is the concentration of sample solution($g \cdot (100\ mL)^{-1}$). Note that the units of the above characters should be unified when calculate the density or concentration.

The instrument used for the measurement of rotation angle is called polarimeter, which is composed of five main working parts: sodium lamp, polarizer(a Nicol prism), sample tube, analyzer(another Nicol prism), eyepiece and vernier scale. The sample is placed into a tube with designated length. The tube should be filled with the sample solution and be screwed tightly. If bubbles are not avoidable, force them to the bulge position on the tube(see Figure 3-5).

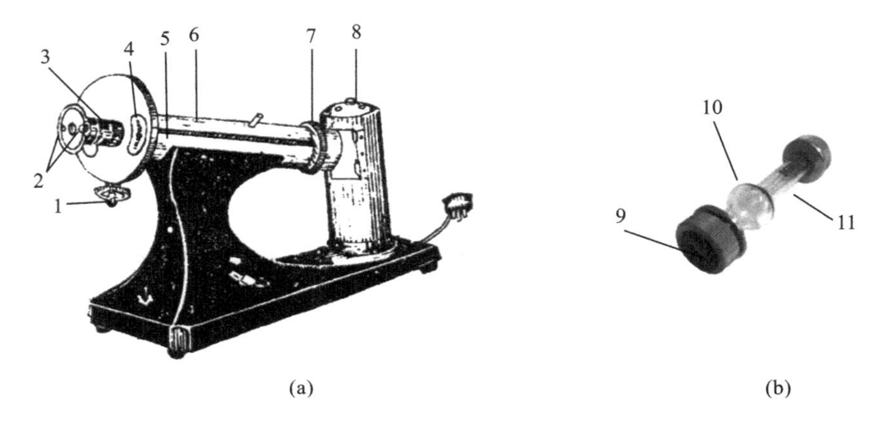

Figure 3-5 The Appearance and Working Parts of WXG-4 Polarimeter and Its Sample Tube

1—hand wheel; 2—magnifier; 3—focal adjustment knob; 4—vernier scale on analyzer; 5—sample tube cradle; 6—sample cell cap; 7—polarizer; 8—sodium lamp; 9—glass plate; 10—bulge for bubbles; 11—sample cell

The instrument should be checked initially by making a zero reading with a sample cell filled only with solvent. If the zero reading does not correspond with the zero-degree calibration mark, the difference in reading must be used to correct all subsequent readings.

Adjust to sharpen the focus first, then rotate the handwheel clockwisely(to the right)or counterclockwisely(to the left) until two fields are clearly visible(often a vertical bar split down the middle of a background field) and interchange quickly in a small rotation range. Then back off a little slowly until no obvious split-image and the entire visual field is as uniform as possible(see Figure 3-6). Read the rotation angle from the analyzer scale and use the vernier scale to estimate the reading to a fraction of a degree(see Figure 3-7). Repeat the determination several times and use the average value to calculate its specific rotation or concentration.

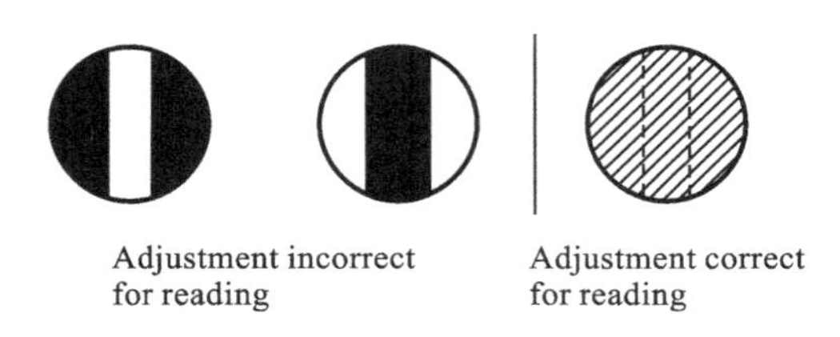

Figure 3-6 Split-fields and Uniform Field for Reading

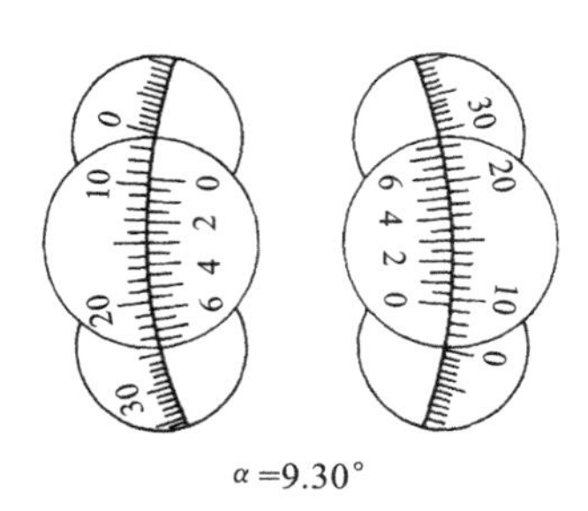

$\alpha=9.30°$

Figure 3-7　A Reading Demonstration on Vernier Scale

Ⅲ Apparatus、Reagents and Materials

【Apparatus】Polarimeter, sample tubes.

【Reagents and materials】Distilled water, glucose solution, fructose solution; lens paper.

【Physical constants】

Compound	H_2O	D-(+)-Glucose	D-(−)-Fructose
Specific Rotation($[\alpha]_D^{20}$)	0	+52.25°	−92.25°

Ⅳ Procedures

(1) Rinse the sample tube with the solvent(H_2O) and fill sample tubes with solvent and sample solution to be analyzed respectively, place tubes in the cradle of polarimeter in turn, and adjust light source.

(2) Throw the power switch of polarimeter to the "ON" position for 5～10 min until the sodium lamp is properly warmed.

(3) Sharpen the focus, rotate analyzer scale to observe the field until the correct uniform field is found.

(4) Record the reading by the vernier scale(see Figure 3-7), repeat step (3) at least twice.

(5) Record the lengths of the tubes used, the temperature and the solvent used(water can be omitted).

(6) Calculate the concentration of glucose or fructose solution according to the measured rotation angle(Note: calibrating reading should be excluded.)and their specific rotation.

Ⅴ Notes and Instructions

【Requirements for preview】

(1) Review the concepts: rotation angle and specific rotation, chiral and achiral, optical activity, enantiomerism.

(2) Get to know the working parts of a polarimeter and its working principle.

【Experimental precautions】

(1) The illumination source should be warmed up for 5～10 min before measuring to get a maximum illumination in the eyepiece.

(2) Be sure no bubbles are in the light-passing way and all stay in the bulge of the tube, or the light would be reflected back by the bubbles and cause an error to the determination. For the same reason, the glass endplates of the sample tube must be clean to avoid reflection of light.

(3) Repeated determinations are very necessary to avoid the visual errors.

(4) The rotation angle of distilled water used for calibration of zero-reading (α_c) must be excluded in the calculation. That's to say, the actual rotation angle of the sample must be: ($\alpha-\alpha_c$).

Ⅵ Post-lab Questions

(1) Why do we need to run a solvent blank measurement for the calibration?

(2) How to determine the rotation angle of some sample solution is $-10°$ but not $+170°$?

Ⅶ Verbs

plane-polarized light 平面偏振光;
Nicol prism 尼科尔棱镜;
polarimeter 旋光仪;
specific rotation 比旋光度;
rotation angle 旋光度;
vernier scale 游标刻度盘;
polarizer 起偏镜;
analyzer 检偏镜;
calibration 校正

实验十六　有机化合物的官能团性质实验

化学性质实验是针对物质分子官能团的性质进行验证和鉴定的实验。根据官能团的性质实验,可以将未知物进行基本的归类,可以为进一步确定分子结构提供必要的信息。

性质实验的装置较为简单,大多数可以在试管中完成,有时需要用加热和气体吸收装置。性质实验要做到安全、准确,需要注意实验操作的细致和准确,否则会得到错误的结果,甚至带来安全隐患。

化学性质实验的注意事项如下。

(1) 试剂准备与储存:性质实验中所用的试剂有气体、液体,也有固体,根据实验的需要,可以自行制备所用的试剂,如乙烯和二氧化碳等气体,或某些必须新鲜制备的试剂(如新制氢氧化铜溶液)。也可以直接购买分析纯化学试剂进行配制,如一些液体和固体等。这些试剂都必须保存在相应的试剂瓶中,贴上标明试剂名称、浓度、配制日期的标签。为方便使用,液体试剂可以取少量储存在带滴头的小试剂瓶中,少量固体可放在干燥器中保存备用。

(2) 试剂的使用:使用任何试剂之前都必须了解该试剂的毒性、腐蚀性以及挥发性,以防出现安全问题。对于强毒性或强腐蚀性的试剂,必须戴上手套和防护眼镜小心取用;对于具有挥发性的试剂,应该在通风橱中取用。打开的试剂瓶必须在取用后立即盖好放回原处,不可错盖或不盖,以免造成试剂的混乱而影响实验的进行和实验结果。试剂必须按照实验需要取用相应的量,以免造成浪费和污染环境,不慎多取或误取的试剂不可倒回原储存瓶中。

(3) 必须按照实验操作步骤进行实验,如试剂的加入顺序、使用浓度和使用量等,不可随意改动。若有不同想法,鼓励学生与教师讨论,经教师同意后进行。实验过程中应该仔细观察实验现象并及时做好记录，对于非预料中的现象应该分析原因，排除影响因素后再进行实验。应该用科学的态度认真对待每一个实验现象，不要轻易下结论或放弃观察。

(4) 实验结束后，将所有试剂瓶按编号放在指定位置，标签朝外，检查是否盖好，固体是否放回干燥器中。使用过的试管等仪器必须清理干净，并将洗好的仪器正确放置在指定地方保存。

实验 16.1　卤代烃的性质

一、实验目的

理解不同结构卤代烃的亲核取代反应活性的差异。

二、实验试剂

饱和硝酸银的乙醇溶液，1-溴丁烷，2-溴丁烷，2-甲基-2-溴丙烷，溴苯，3 mmol·L^{-1} HNO_3溶液，1-氯丁烷，1-碘丁烷。

三、实验内容

(1) 将 4 支干燥的试管依次标上记号并各加入 1 mL 饱和硝酸银乙醇溶液[1]，然后分别加入 2 滴 1-溴丁烷、2-溴丁烷、2-甲基-2-溴丙烷、溴苯，振摇后注意观察出现沉淀的先后次序。若不见沉淀析出，可在 70 ℃左右热水浴中加热，3 min 后观察现象，在每支有沉淀的试管中各滴入 1 滴 3 mmol·L^{-1} HNO_3溶液，振摇，若沉淀溶解则不是卤化银。根据实验现象，排列出四种溴代物的反应活性次序，说明烃基结构不同对反应速率的影响。

(2) 在 3 支干燥试管中各加入 1 mL 饱和硝酸银乙醇溶液，然后分别加入 2 滴 1-氯丁烷、1-溴丁烷、1-碘丁烷，充分振摇后，观察沉淀析出的先后次序。若未见沉淀析出，可在 70 ℃左右水浴中加热。根据实验结果，排列出三种卤代丁烷的反应活性次序，说明不同卤原子对反应速率的影响。

四、注解

[1] 在 18～20 ℃时，硝酸银在无水乙醇中的溶解度为 2.1 g·$(100\ mL)^{-1}$，由于卤代烃能溶于乙醇而不溶于水，所以用乙醇做溶剂，能使反应处于均相，有利于反应顺利进行。

五、思考题

伯、仲、叔卤代烷与硝酸银乙醇溶液作用的活性次序有何不同？试说明原因。

实验 16.2　醇和酚的性质

一、实验目的

掌握一元醇和多元醇性质的差异，以及酚类化合物的检验方法。

二、实验试剂

正丁醇，仲丁醇，叔丁醇，苯甲醇，卢卡斯试剂，浓盐酸，10%乙二醇，10%丙三醇(甘油)，

2% $CuSO_4$溶液，10% NaOH 溶液，1%苯酚溶液，1%α-萘酚溶液，1%间苯二酚溶液，饱和溴水，1%$FeCl_3$溶液。

三、实验内容

1. 醇的性质

(1) 醇的取代：与卢卡斯(Lucas)试剂反应[1]——伯、仲、叔醇的鉴别。

在 4 支试管中分别加入 2 mL 卢卡斯试剂，再分别加入 5～6 滴样品正丁醇、仲丁醇、叔丁醇、苯甲醇，塞住试管，用力振摇数分钟，然后将试管放入 25～30 ℃的水浴中，观察其变化。记录各试管中产生混浊和分层的时间，解释其现象差异。

用 2 mL 浓盐酸代替卢卡斯试剂，按照上述方法进行实验，比较结果。

(2) 多元醇的氢氧化铜实验。

取 6 滴 2%$CuSO_4$溶液，滴加 10%NaOH 溶液至氢氧化铜沉淀全部析出，边振荡边滴加样品乙醇、10%乙二醇、10%丙三醇各 2～3 滴，观察结果并加以比较。

2. 酚的性质

(1) 与溴水的反应[2]。

取 1%苯酚溶液 5 滴，再逐滴加入饱和溴水，边加边摇，观察现象并解释。

(2) 与 $FeCl_3$的颜色反应[3]。

分别加入 5 滴 1%苯酚溶液、1%α-萘酚溶液、1%间苯二酚溶液，然后在各管中加入 1 滴 1%$FeCl_3$溶液，观察各管呈现的颜色。

四、注解

[1] 卢卡斯试剂是浓盐酸与无水氯化锌的混合物，又称盐酸-氯化锌试剂，在有机分析中用作伯、仲、叔醇的鉴别试剂。本实验仅适用于在卢卡斯试剂中能溶解的醇，通常只用于鉴别 C_3～C_6 醇，因为 C_1～C_2 醇反应后所生成的氯代烷是气体，而大于 6 个碳的醇不溶于卢卡斯试剂故不适用。用浓盐酸代替卢卡斯试剂，只有叔醇起反应。

[2] 室温下生成的 2,4,6-三溴苯酚在水中的溶解度为 0.007 g·$(100\ mL)^{-1}$，故呈白色混浊。但过量的溴水也可将 2,4,6-三溴苯酚氧化成淡黄色的 2,4,4,6-四溴环己二烯酮，所以溴水不能加过量。

OH, Br, Br, Br　$\xrightarrow{Br_2/H_2O}$　O, Br, Br, Br, Br　+HBr

[3] 大多数酚类和浓度较高的烯醇类物质都能与 $FeCl_3$反应，生成有色配合物。$FeCl_3$既是显色剂，又是氧化剂，过量的 $FeCl_3$溶液能与某些酚起氧化反应，例如，对苯二酚被氧化成对苯醌，α-萘酚被氧化为溶解度很小的联苯酚，并以白色沉淀析出。

实验 16.3　醛、酮的性质

一、实验目的

(1) 掌握羰基化合物的化学特性以及鉴定方法。

(2) 学习用简单的化学方法对未知物进行鉴定。

二、实验试剂

2,4-二硝基苯肼试剂，乙醛，丙酮，95%乙醇，碘试剂，10%NaOH 溶液，菲林试剂(A)和菲林试剂(B)。

三、实验内容

1. 羰基的性质

(1) 与 2,4-二硝基苯肼作用：取 2 支试管，各加入 15 滴 2,4-二硝基苯肼试剂，再分别加入 2～3 滴乙醛、丙酮，振荡，观察有无沉淀生成，并注意结晶的颜色。

(2) 碘仿反应：取 3 支试管，分别加入 8～10 滴乙醛、丙酮、95%乙醇，再加入 10～15 滴碘试剂，摇匀后，滴加 10%NaOH 溶液至碘的颜色刚好消失为止，观察现象，并嗅其气味。

2. 醛基的性质

与菲林试剂反应：在 1 支试管中先后加入 1 mL 菲林试剂(A)和菲林(B)，混合均匀，并平分到 3 支试管中，再分别加入 5 滴甲醛、乙醛和丙酮，摇匀，放于沸水浴中加热 5 min，观察现象，并比较结果。

3. 辨别未知物

有 A、B、C、D、E 五瓶液体，已知它们是甲醇、乙醇、正丙醇、乙醛或丙酮，但不知各瓶中装的是哪一种。请设计一个辨别方案并用实验证实 A～E 瓶中各装的是什么物质。

四、思考题

(1) 哪些试剂可用以区别醛类和酮类?

(2) 什么是碘仿反应? 哪种丁醇能起碘仿反应?

实验 16.4　羧酸及取代羧酸的性质

一、实验目的

理解有机酸的酸性和取代羧酸的官能团特性。

二、实验试剂

10% 甲酸，10% 乙酸，10% 草酸，苯甲酸，10% Na_2CO_3溶液，10% NaOH 溶液，10%盐酸，饱和苯甲酸溶液，饱和水杨酸溶液，1% $FeCl_3$溶液，水杨酸固体，饱和澄清石灰水，10% 乙酰乙酸乙酯。

三、实验内容

1. 酸性

(1) 在3支小试管中各加2 mL 10% Na_2CO_3溶液,再分别加入5滴10%甲酸、10%乙酸、10%草酸,观察是否有气体逸出现象。并解释。

(2) 在1支小试管中加入0.1 g固体苯甲酸,加1 mL蒸馏水,振摇,观察是否溶解。再加入数滴10% NaOH溶液,振摇,观察现象。然后加入10%盐酸数滴,观察又有何现象,说明原因。

2. 取代羧酸的性质

(1) 与$FeCl_3$反应:取2支试管,分别加入5滴饱和苯甲酸溶液、饱和水杨酸溶液,再各加1～2滴1% $FeCl_3$溶液。观察各试管有何现象。

(2) 水杨酸脱羧:取1支干燥试管,加入约0.5 g水杨酸固体,塞上带玻璃弯导管的塞子,玻璃导管的另一端插入盛有饱和澄清石灰水的试管中。加热水杨酸(试管口稍微向上倾斜),使它熔化并继续加热至沸。观察试管中有何变化。

(3) 酮式-烯醇式互变现象:在1支试管中加入1 mL 10%乙酰乙酸乙酯,加入1滴1% $FeCl_3$溶液,观察颜色变化。向此溶液快速滴入数滴饱和溴水,观察颜色变化。说明这些颜色变化的原因。

四、思考题

设计一个实验,证明酸性次序为:苯甲酸>碳酸>苯酚>苯甲醇。

实验16.5　糖类物质的性质

一、实验目的

掌握实验室检验单糖和多糖的方法。

二、实验试剂

班乃德(Benedict)试剂,2%葡萄糖溶液,2%果糖溶液,2%乳糖溶液,2%蔗糖溶液,2%麦芽糖溶液,2%淀粉溶液,盐酸苯肼-乙酸钠,1% 碘液,浓硫酸。

三、实验内容

1. 班乃德实验

取5支试管,各加入10滴班乃德试剂,再分别加入5滴2%葡萄糖溶液、2%果糖溶液、2%蔗糖溶液、2%麦芽糖溶液、2%淀粉溶液,在沸水浴中煮沸2～3 min,取出冷却,观察有无红色沉淀,比较其结果。

2. 糖脎的生成

取3支试管,分别加入1 mL 2%葡萄糖溶液、2%乳糖溶液、2%麦芽糖溶液,再各加入0.5 mL盐酸苯肼-乙酸钠试剂。将3支试管充分振摇,置沸水浴中加热30 min后取出,若无结晶,可将试管放入冷水中冷却后再观察。比较各试管中成脎的速度和颜色。取各种糖脎少许，在显微镜下观察糖脎的晶型(见图 3-8)。

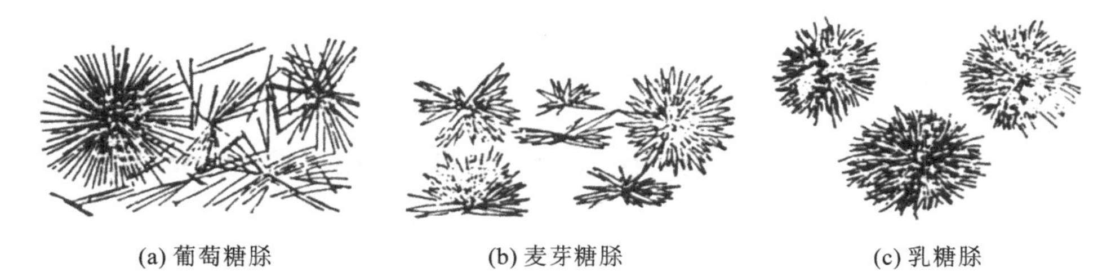

(a) 葡萄糖脎　(b) 麦芽糖脎　(c) 乳糖脎

图 3-8　糖脎的晶型

3. 淀粉的水解及与碘的反应

在 1 支试管中加入 3 mL 2%淀粉溶液，再加 5～6 滴浓硫酸。将洁净的表面皿上滴几滴 1%碘液在不同的位置。将试管于沸水浴中加热，每隔 30 s 取 1 滴水解液滴在表面皿上与碘液混合，观察颜色变化。解释实验现象。

在 1 支试管中加入 5 滴 2% 淀粉溶液，加 1 mL 水，然后加入 1 滴 1% 碘液，摇匀，观察有什么变化。将溶液加热，有何现象？放冷后又有什么变化？解释实验现象。

实验 16.6　氨基酸和蛋白质的性质

一、实验目的

了解氨基酸和蛋白质的基本化学特性及检验方法。

二、实验试剂

1% 甘氨酸溶液，1%谷氨酸溶液，1%酪氨酸溶液，清蛋白溶液，0.2%茚三酮溶液，1% $CuSO_4$溶液，1% HCl 溶液，1% NaOH 溶液，10% NaOH 溶液，$(NH_4)_2SO_4$(固体)，浓盐酸，红色石蕊试纸。

三、实验内容

1. 氨基酸和蛋白质的水合茚三酮反应

取 4 支试管，分别加入 1 mL 1%甘氨酸溶液、1%谷氨酸溶液、1%酪氨酸溶液和清蛋白溶液，再分别加入 3～4 滴 0.2%茚三酮溶液，摇匀，在沸水浴中加热 10～15 min，取出，观察颜色变化。

2. 蛋白质的两性反应

取 1 支试管，加入 2 mL 清蛋白溶液，逐滴加入 1% HCl 溶液，边加边摇动试管，观察有无沉淀产生。沉淀出现后继续滴加 1% HCl 溶液，观察有何现象。改用 1% NaOH 溶液如上操作，有何现象？用红色石蕊试纸测定沉淀出现时溶液的酸碱性。

3. 蛋白质的盐析

取 1 支试管，加入 1 mL 清蛋白溶液，然后加入固体$(NH_4)_2SO_4$，边加边振荡，待加到一定量时，观察有无沉淀产生。用水稀释后，又有何现象？

4. 蛋白质的变性

在1支试管中加入5滴浓盐酸,将试管倾斜,小心地沿着试管壁滴加5滴清蛋白溶液,观察两液层接触界面上的现象。

在1支试管中加入5滴清蛋白溶液及2滴1%$CuSO_4$溶液,观察有何现象。

Experiment 16 Properties of Organic Compounds

Experiment 16.1 Properties of Alcohols, Phenols, Aldehydes, Ketones, Carboxylic Acids and Their Derivatives

Ⅰ Objectives

(1) To understand reaction characters on main functional groups of major organic compounds, such as halohydrocarbons, alcohols, phenols, aldehydes, ketones, carboxylic acids and their derivatives.

(2) To learn how to determine the class of compounds to which the unknown belongs.

Ⅱ Apparatus and Reagents

【Apparatus】 Test tubes, test tube clips, capillary dropping tube, water bath[1].

【Reagents】 Saturated ethanol solution of $AgNO_3$, 1-bromobutane, 2-bromobutane, 2-methyl-2-bromopropane, bromobenzene, 3 mmol · L^{-1} HNO_3, 2% $CuSO_4$, 10% NaOH, 10% glycerol, 10% 1,2-ethanediol, 95% ethanol, 1% phenol, 1% α-naphthol, 1% resorcinol, 1% $FeCl_3$, 2,4-dinitrophenylhydrazine, ethanol (CH_3CH_2OH), acetone (CH_3COCH_3), formaldehyde (HCHO), acetaldehyde (CH_3CHO), Fehling reagent (A) and (B), iodine reagent, salicylic acid (s), 10% ethyl acetoacetate, saturated bromine water, saturated lime water, 10% Na_2CO_3, 10% formic acid (HCOOH), 10% acetic acid (CH_3COOH), 10% oxalic acid (HOOCCOOH), benzoic acid (s), 10% HCl.

Ⅲ Procedures

1. Properties of halohydrocarbons—Reaction with $AgNO_3$

Take 4 dry tubes and label in sequence. Add 1 mL of saturated ethanol solution of $AgNO_3$ in each tube, and then add 2 drops of 1-bromobutane in the first tube, 2 drops of 2-bromobutane in the second, 2 drops of 2-methyl-2-bromopropane in the third and 2 drops of bromobenzene in the last. Observe if there are precipitates (Note: AgBr is yellow precipitate) appearing in sequence during shaking, otherwise, place the tubes in warm water (70 ℃) for 3 min. Observe the formation of precipitates. Then remove water bath and cool, add 1 drop of 3 mmol · L^{-1} HNO_3 into each tube to see if the precipitates dissolved (Note: AgBr can not dissolve in strong acids). According to the results, rank the relativities of the above four halohydrocarbons in the reaction with $AgNO_3$ and give an explanation.

2. Properties of polyhydric alcohols—Reaction with $Cu(OH)_2$

Take 3 test tubes, add 6 drops of 2% $CuSO_4$ solution respectively, then add about 4 drops of 10% NaOH in each tube, the precipitate are formed immediately by shaking. And then add 3 drops of 10% glycerol into the first tube, 3 drops of 10% 1,2-ethanediol into the second and 3 drops of 95% ethanol into the third. Shake the mixture and then let them stand for a while. Observe the phenomenon of the mixture respectively.

3. Properties of phenols—Reaction with $FeCl_3$

Take 3 test tubes, add 5 drops of 1% phenol, 1% α-naphthol, 1% resorcinol respectively, then place 1 drop of 1% $FeCl_3$ in each tube. Observe the changes shown in every tube.

4. Properties of aldehydes and ketones—Reaction with 2,4-dinitrophenylhydrazine

Place 15 drops of 2,4-dinitrophenylhydrazine in each of two test tubes, then add 5 drops of acetone, ethanol respectively in each tube. Shake them vigorously and then stand for a while. Observe if the precipitate is formed in each tube. However, if a precipitate does not form immediately, let the solution stand for 30 s in warm water. And then cool it in the room temperature after shaking it vigorously. Observe the change.

5. Iodoform test

Take 3 test tubes, add 5 drops of acetaldehyde, acetone, 95% ethanol respectively, and then add 10～15 drops of iodine reagent in each tube. Then, add 10% NaOH in each tube drop by drop with shaking until the brown color of the solution disappears. Observe whether the yellow iodoform is formed.

6. Oxidation reaction of aldehydes with Fehling reagent

Mix 1 mL of Fehling reagent(A) with 1 mL of Fehling reagent (B) in a test tube, and then pour the mixture into three tubes equally. And then do the following: Add 5 drops of formaldehyde in the first tube, 5 drops of acetaldehyde in the second and 5 drops of acetone in the third. Shake these tubes and then stand them into boiling water for 5 min. Take them out and observe the changes.

7. Properties of carboxylic acids and their derivatives

1) Acidity of carboxylic acids

(1) Add 2 mL of 10% $NaCO_3$ into 3 clean test tubes, then add 5 drops of the following 3 acids into 3 tubes respectively: 10% formic acid(HCOOH), 10% acetic acid(CH_3COOH), 10% oxalic acid(HOOCCOOH). Observe if there is gas giving off, explain.

(2) Place 0.1 g of solid benzoic acid into a test tube, and then add 1 mL of distilled water. Shake it to see if the solid dissolved. Then, add several drops of 10% NaOH, shake and observe the change. Then, add several drops of 10% HCl into the tube to see what will happen. Explain all the changes.

2) Decarboxylation of salicylic acid by heating

Take 0.2 g of salicylic acid solid into a dry test tube. Stopper the test tube with a rubber stopper that joints a bend dropper. Hold the tube by a test tube clip and heat it(the mouth of the tube slants down gently)on a flame until the oxalic acid melts. At the same time, let the dropper plug in a tube that fills the saturated solution of lime and water. Observe the change that takes place in the saturated solution of lime and water [2].

3) Reaction of salicylic acid with $FeCl_3$

Place a little salicylic acid solid into a test tube and then add 1 mL of water. Shake the tube, and add again 1～2 drops of 1% $FeCl_3$. Observe the color change of the solution. Explain the change.

4) Keto-enol tautomerism of ethyl acetoacetate

Place 5 drops of ethyl acetoacetate and 2 mL of water in a test tube, and then add 1～2 drops of 1% $FeCl_3$. Observe the color change of the mixture. Then add 3～4 drops of saturated bromine water, what phenomenon do you observe? And what change has taken place after waiting a while? Please explain for the series of changes.

8. Identification of unknowns

Now there are 5 bottles of liquids(labeled A, B, C, D and E)on the table. We just know these liquid compounds are: methanol(CH_3OH), ethanol(CH_3CH_2OH), propanol($CH_3CH_2CH_2OH$), acetaldehyde(CH_3CHO)and acetone(CH_3COCH_3), but we don't know exactly which bottle contains which compound. Please design a reasonable and simple scheme and carry out some corresponding tests to identify them one by one.

Ⅳ Notes and Instructions

【Notes】

[1] Heating a bottle of water by using a 250 mL beaker as a water bath.

[2] The test tube filled with solution of lime and water has been removed away first and then the flame is removed after completing the experiment.

【Requirements for preview】

(1) To review the main reaction characters of R—OH, ph—OH, —CHO, $\rangle C{=}O$ and —COOH.

(2) To get an idea of identification of unknown compounds according to their different reaction characters of functional groups.

【Experimental precautions】

(1) All the test tubes must be cleaned carefully before you start a new experiment.

(2) Remember to cover the reagent bottle immediately after you take out the reagent. Don't mix the stopper of different bottles in case of the reagents being polluted.

Ⅴ Post-lab Questions

How to verify the acidity of benzoic acid, carbonic acid, phenol and benzyl alcohol is in decreasing order? Please design an experiment to prove this conclusion.

Ⅵ Verbs

halohydrocarbon 卤代烃；
Fehling reagent 菲林试剂；
acetaldehyde 乙醛；
glycerol 甘油，丙三醇；
ethanediol 乙二醇；
naphthol 萘酚；
resorcinol 间苯二酚；
formaldehyde 甲醛，蚁醛；
2,4-dinitrophenylhydrazine 2,4-二硝基苯肼；
oxalic acid 草酸；
salicylic acid 水杨酸；
ethyl acetoacetate 乙酰乙酸乙酯；
iodoform 三碘甲烷，碘仿；
lime water 石灰水；
tautomerism 互变现象

Experiment 16.2 Properties of Saccharides, Amino Acids and Proteins

Ⅰ Objectives

(1) To enhance the understanding on properties of carbohydrates, amino acids and proteins.

(2) To learn how to identify the carbohydrates, amino acids and proteins.

Ⅱ Apparatus and Reagents

【Apparatus】 Test tubes, capillary dropping tube, water bath.

【Reagents】 Benedict's reagent, 2% D-glucose, 2% D-fructose, 2% sucrose, 2% maltose, 2% lactose, 2% starch, mixture solution of phenylhydrazine hydrochloride and sodium acetate, 1% iodine, 1% glycine, 1% glutamic acid, 1% lysine, 0.2% ninhydrin, 5% NaOH, 1% $CuSO_4$, egg albumin solution, 0.5% $Pb(Ac)_2$.

Ⅲ Experiments

1. Oxidation of saccharides—Benedict's test [1]

Take five tubes. Place 15 drops of Benedict's reagent into each tube, and then add 5 drops of the following 2% saccharide solutions in separate tubes of the five: glucose, fructose, sucrose (it should be pure), maltose and starch. Put the tubes in a boiling water bath for 3 min. Remove and cool them, and observe the results.

2. Reaction with phenylhydrazine—Formation of osazones

Take 3 test tubes. Add 1 mL of 2% glucose, 2% lactose, 2% maltose in turn, and then add 0.5 mL of mixture solution of phenylhydrazine hydrochloride and sodium acetate in each tube. Shake the tubes to fully mix the solution, then heat them in a boiling water bath for 30 min. Take them out and cool for a while to allow the reaction products crystalize from the solutions. Observe the formations of osazones of different saccharides. Compare the crystalling rate and color between 3 kinds of osazones. And drop a little of osazones of each on glass plates to observe their crystal forms under a microscope(see Figure 3-8).

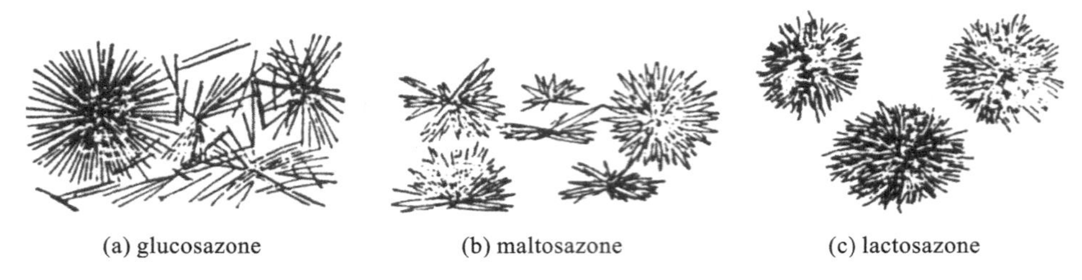

(a) glucosazone　(b) maltosazone　(c) lactosazone

Figure 3-8　Crystal Form of Osazones

3. Reaction of polysaccharides with iodine

Take two tubes. Place 1 drop of 2% starch in the tube, then dilute it with 3 mL of distilled water. Add 1 drop of 1% iodine to each test tube with shaking and observe the results. Put the tubes into boiling water bath for 5～10 min and note the change. Remove and cool them, what changes do you observe ? Explain.

4. Colored reaction of amino acids and proteins—Reaction with ninhydrin

Add 10 drops of protein solution: 1% glycine, 1% glutamic acid, 1% tyrosine in 3 test tubes, and add 3 drops of 0. 2% ninhydrin into each tube with shaking, then put them in a boiling water bath for 5～10 min. Observe the results and explain.

5. Denaturation of proteins—Reaction with strong acid or heavy metal salts [2]

Place 5 drops of concentrated HCl in a tube, then tilt the tube and add 5 drops of egg albumin solution carefully drop by drop without shaking. Observe what happened at the interface of two layers.

Place 5 drops of egg albumin solution into another tube, then add 2 drops of 1% $CuSO_4$. Shake the tubes and observe the results. Explain.

Ⅳ Notes

[1] For many years the Benedict test was preliminary screen for diabetes because the precipitate is formed if excess glucose is presented in the urine. The color of precipitates can be changed with green, yellow and red according to the amount of glucose present. But because all common aldoses and ketoses give positive reducing sugar tests, more specific tests have been devised to identify the presence of glucose specifically in body fluids as evidence of diabetes, rather than other sugars excreted due to some other form of abnormal metabolism such as fructosemia. Proteins known as enzymes are usually extremely specific in reacting with compounds. The enzyme glucose oxidase has been mixed with dyes and placed on a paper strip(test-paper) so that, when dipped into urine, it will record the presence and relative amount of glucose present. This is the basis of the widely used commercial test kit that both takes a sample of blood by pricking the finger and analyzes it.

[2] Heavy metal ions, such as silver, lead and mercury, precipitate proteins by combination of a metal cation with the free carboxyl groups of the protein. The antiseptic action of silver nitrate and mercuric chloride depends upon the precipitation of the proteins

present in bacteria. But the precipitate will be dissolved if add excess heavy metal ions in the precipitate solution.

Ⅴ Post-lab Questions

(1) Why is egg white or milk used as an antidote to treat patients who have swallowed heavy metal salts, such as silver nitrate or mercuric chloride?

(2) Why is alcohol used to disinfect areas of the skin prior to surgery?

Ⅵ Verbs

saccharide 糖类；
glucose 葡萄糖；
fructose 果糖；
sucrose 蔗糖；
maltose 麦芽糖；
osazone 糖脎；
glucosazone 葡萄糖脎；
maltosazone 麦芽糖脎；
lactosazone 乳糖脎；
glycine 甘氨酸；
glutamic acid 谷氨酸；
tyrosine 酪氨酸；
ninhydrin (水合)茚三酮；
albumin 清蛋白，白蛋白

第四部分　有机化合物的合成

Part 4　Synthesis of Organic Compounds

实验十七　正溴丁烷的制备

一、实验目的

(1) 理解亲核取代反应的机理，学习制备卤代烃的原理和方法。

(2) 学习正确使用浓硫酸。

(3) 掌握带有有害气体吸收装置的回流和加热操作方法。

(4) 了解分离提纯液体化合物的一般方法和步骤，巩固分液漏斗的操作。

二、实验原理

伯卤代烃正溴丁烷是通过正丁醇与溴化钠及浓硫酸反应制备的。

主要反应：

$$NaBr + H_2SO_4 \longrightarrow HBr + NaHSO_4$$

$$n\text{-}C_4H_9OH + HBr \rightleftharpoons n\text{-}C_4H_9Br + H_2O$$

副反应：

$$CH_3CH_2CH_2CH_2OH \xrightarrow[\triangle]{\text{浓 } H_2SO_4} CH_3CH_2CH{=}CH_2 + H_2O$$

$$2CH_3CH_2CH_2CH_2OH \xrightarrow[\triangle]{\text{浓 } H_2SO_4} n\text{-}C_4H_9OC_4H_9\text{-}n + H_2O$$

反应副产物主要是丁醚、丁烯等。在正溴丁烷的分离纯化过程中，用硫酸洗涤粗产品以洗去未反应的正丁醇，并除去反应副产物烯烃和醚；残余的酸用碳酸氢钠中和后，粗产物用无水氯化钙进行干燥，以除去残留的正丁醇。

三、仪器与试剂

【仪器】圆底烧瓶(25 mL、10 mL)，球形冷凝管，分液漏斗(50 mL×2)，锥形瓶(50 mL、25 mL、10 mL)，常压蒸馏装置，气体吸收装置，折光仪。

【试剂】正丁醇，浓硫酸，溴化钠(固)，饱和碳酸氢钠溶液，5% 氢氧化钠溶液，无水氯化钙。

【物理常数】

化合物名称	相对分子质量	密度/(g·cm^{-3})	熔点/℃	沸点/℃	折光率(n_D^{20})	溶解度/[g·(100 mL)$^{-1}$]		
						水	乙醇	乙醚
正丁醇	74.12	0.8097	−89.5	117.7	1.3993	7.4	∞	∞
正溴丁烷	137.02	1.2758	−112.4	101.6	1.4399	i	s	s
溴化钠	102.89	3.200	755	1390	—	90(vs)	6	i
浓硫酸	98.08	1.8318	10.4	335.5	—	∞	∞	—

注：i 表示不溶；s 表示溶解；vs 表示易溶；∞表示混溶。

【试剂用量】

试　剂	用　量
蒸馏水	3 mL
浓硫酸	4 mL(7.3 g,0.074 mol)
正丁醇	3.0 mL(2.4 g,0.033 mol)
溴化钠	4.0 g(0.039 mol)

四、实验步骤

1. 安装实验装置

在 25 mL 圆底烧瓶上安装回流冷凝管，冷凝管的上口接气体吸收装置(见图 4-1)，用 5% 氢氧化钠溶液做吸收液，以除去反应中生成的溴化氢气体，倒置的漏斗边缘恰好触及烧杯中的稀氢氧化钠溶液表面。

2. 加入反应物

在圆底烧瓶中加入 3 mL 水，慢慢边振摇边加入 4 mL 浓硫酸，混合均匀后冷至室温。再依次加入 3.0 mL 正丁醇和 4.0 g 溴化钠。充分振摇后，加入几粒沸石。将反应瓶放置在电热套上，用夹子夹好，连接回流冷凝管和吸收溴化氢气体的装置，注意勿使漏斗全部埋入水中，以免倒吸。

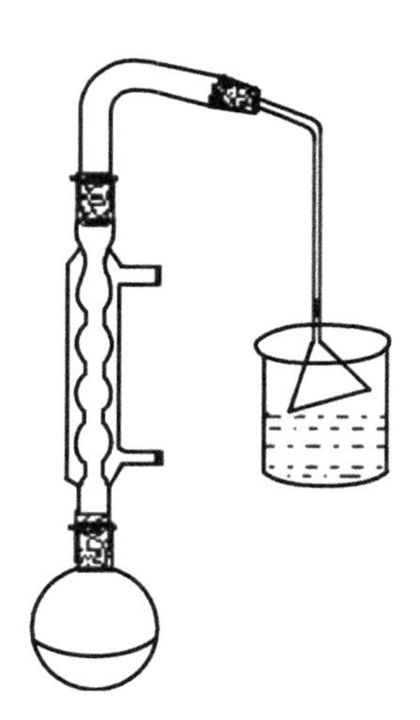

图 4-1　正溴丁烷制备反应装置

3. 回流

将烧瓶在电热套中加热至沸，平稳回流并不时加以摇动，使反应完全。此时反应瓶内溶液分层，上层即为正溴丁烷。回流 30 min 使溴化钠反应完全，停止加热，稍微冷却反应装置(直至不太烫手)。

4. 蒸馏

移去回流冷凝管，再向烧瓶内加两粒沸石，改为常压蒸馏装置(将蒸馏头置于圆底烧瓶上，连接直形冷凝管、尾接管和接收瓶)。加热蒸出粗产物正溴丁烷，直至馏出液透明为止[1]。

5. 萃取

将馏出液移至 50 mL 分液漏斗中，加入等体积的水洗涤[2]。振摇，放气，静置分层，产物位于下层。将产物分出，再用等体积的浓硫酸洗涤[3]。尽量分出下层的硫酸层，有机相再依次用等体积的水和饱和碳酸氢钠溶液洗涤，注意振摇放气。最后一次洗涤时，应尽量将水层分离干净。正溴丁烷分离纯化流程如下：

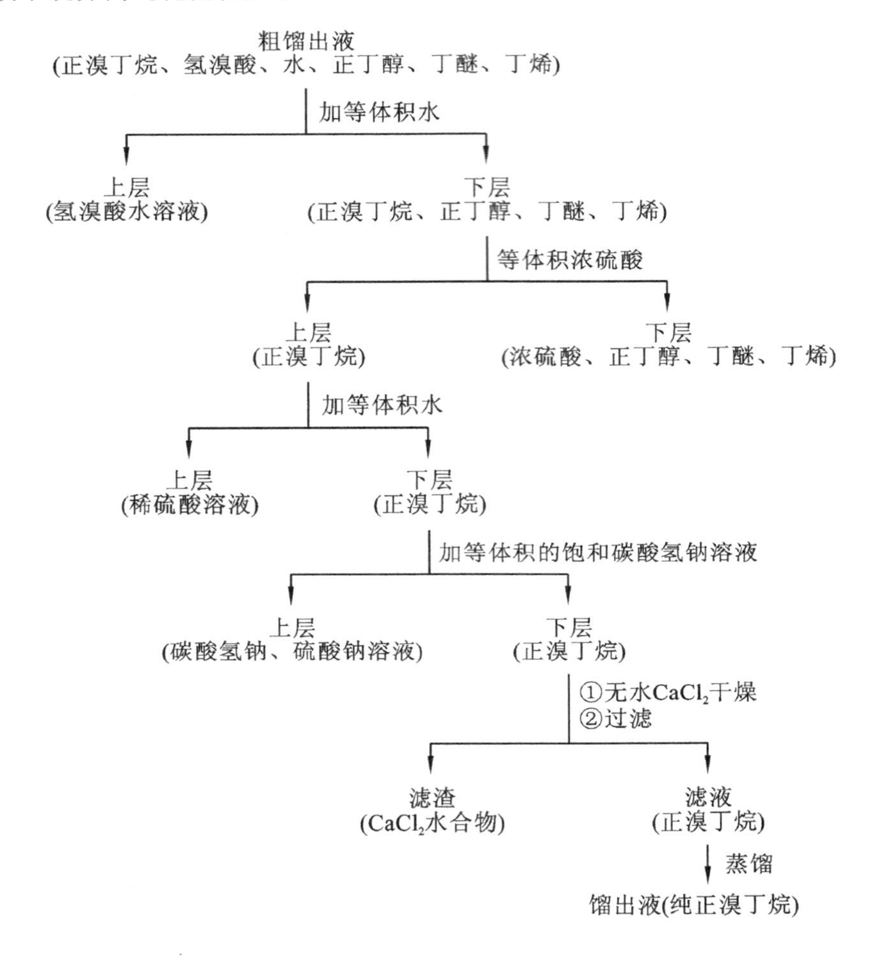

6. 干燥

将有机相转入 25 mL 锥形瓶中，加入 0.5 g 左右无水氯化钙干燥 30 min[4]。间歇摇动锥形瓶，直至液体清亮为止。

7. 蒸馏(精制)

将干燥后的产物过滤到 10 mL 蒸馏瓶中，加入沸石后在电热套上加热蒸馏，收集 98～102 ℃的馏分。

8. 产率测定

称重，产量为 1.8～2.5 g(产率为 40%～55%)。用折光仪测定产物的折光率。纯正溴丁烷为无色透明液体。

本实验约需 5 h。

五、注解和实验指导

【注解】

[1] 正溴丁烷是否蒸完,可从以下三方面判断:

① 馏出液是否由混浊变为澄清;

② 反应瓶上层油层是否消失;

③ 取一支已盛有水的小试管,收集几滴馏出液,振摇,观察有无油珠出现。如无,表示馏出液中已无有机物,蒸馏已完成。

[2] 用水洗涤后馏出液如呈红色,表示存在游离溴,可加入几毫升饱和亚硫酸氢钠溶液洗涤除去。

$$2HBr + H_2SO_4 \longrightarrow Br_2 + SO_2\uparrow + 2H_2O$$

$$Br_2 + 3NaHSO_3 \longrightarrow 2NaBr + NaHSO_4 + 2SO_2\uparrow + H_2O$$

[3] 用浓硫酸洗去粗产物中少量未反应的正丁醇和副产物丁醚等杂质,防止在以后的蒸馏中,正丁醇和正溴丁烷形成共沸物(沸点为 98.6 ℃,含正丁醇 13%)而难以除去。

[4] 无水氯化钙干燥时,不仅可除去水分,还可除去醇类,因为醇类化合物(特别是低级醇)可与氯化钙作用生成结晶醇而不溶于有机溶剂。

【预习要求】

(1) 复习由醇制备卤代烃的反应,理解 S_N2 反应机理。

(2) 分析影响本次实验反应速率和产率的主要因素。

(3) 了解气体吸收装置的种类和使用方法。

(4) 熟悉粗产物正溴丁烷分离纯化流程,确定每一步分离纯化的目的和产物所在的液层。

(5) 了解干燥剂的选择和使用(见本书第二部分实验二)。

【操作注意事项】

(1) 浓硫酸有腐蚀性,易灼伤,小心使用。

(2) 有有害气体放出时,常用气体吸收装置。微型化实验产生的气体量少,可直接用水吸收,但要注意防止倒吸。

(3) 反应物加料顺序不能颠倒。应先加水,再加浓硫酸,然后依次是醇、溴化钠。加水的目的如下:

① 减少 HBr 的挥发;

② 防止产生泡沫;

③ 降低浓硫酸的浓度,以减少副产物的生成。

(4) 在加料过程中及回流时应不时摇动,否则将影响产量。

① 水中加浓硫酸时振摇:防止局部过热;

② 加正丁醇时混匀:防止局部炭化;

③ 加 NaBr 时振摇:防止 NaBr 结块,影响 HBr 的生成;

④ 回流过程中振摇:因为是两相反应,通过振摇增大反应物分子间的接触概率。

(5) 蒸出粗产物后,应趁热倒出残留液,否则结块难以倒出。

六、思考题

(1) 用浓硫酸萃取洗涤时为何要用干燥的分液漏斗?

(2) 写出无水氯化钙吸水后所起化学变化的反应式。为什么蒸馏前一定要将它过滤掉?

(3) 蒸馏出的馏出液中正溴丁烷通常应在下层,但有时出现在上层,为什么?若遇此现象如何处理?

Experiment 17 Synthesis of *n*-Butyl Bromide

Ⅰ Objectives

(1) To study the principle of nucleophilic substitution reactions and preparation of halohydrocarbon.

(2) To know the correct treatment with concentrated sulfuric acid.

(3) To master the technique of reflux with a gas trap apparatus.

(4) To learn the general isolation and purification procedures of a liquid mixture. Be familiar with the operation of a separatory funnel.

Ⅱ Principle

A primary alkyl halide *n*-butyl bromide is prepared easily by allowing *n*-butyl alcohol to react with sodium bromide in the presence of sulfuric acid.

Main reactions:

$$NaBr + H_2SO_4 \longrightarrow HBr + NaHSO_4$$

$$n\text{-}C_4H_9OH + HBr \rightleftharpoons n\text{-}C_4H_9Br + H_2O$$

Side reactions:

$$CH_3CH_2CH_2CH_2OH \xrightarrow[\triangle]{\text{conc. } H_2SO_4} CH_3CH_2CH{=}CH_2 + H_2O$$

$$2CH_3CH_2CH_2CH_2OH \xrightarrow[\triangle]{\text{conc. } H_2SO_4} n\text{-}C_4H_9OC_4H_9\text{-}n + H_2O$$

In the experiment, byproducts are dibutyl ether, butene and so on. During the isolation of the *n*-butyl bromide, the crude product is washed with concentrated sulfuric acid to remove unreacted *n*-butyl alcohol, alkene and ether. Any remaining acid is neutralized by sodium bicarbonate. And then dry the crude product with anhydrous calcium chloride to get rid of residual *n*-butyl alcohol.

Ⅲ Apparatus and Reagents

【Apparatus】 Round-bottom flask(25 mL, 10 mL), Allihn condenser, separatory funnel (50 mL×2), Erlenmeyer flask(50 mL, 25 mL, 10 mL), simple distillation apparatus, gas trap apparatus, refractometer.

【Reagents】 *n*-Butyl alcohol, concentrated sulfuric acid, sodium bromide, saturated aqueous sodium bicarbonate, 5% sodium hydroxide(aq), anhydrous calcium chloride.

【Physical constants】

Compound	M_w	ρ /(g · cm^{-3})	m. p. /℃	b. p. /℃	Refractive Index(n_D^{20})	Solubility/[g · (100 mL)$^{-1}$]		
						Water	Alcohol	Diethyl ether
n-Butyl alcohol	74.12	0.8097	−89.5	117.7	1.3993	7.4	∞	∞
n-Butyl bromide	137.02	1.2758	−112.4	101.6	1.4399	i	s	s
Sodium bromide	102.89	3.200	755	1390	—	90(vs)	6	i
Concentrated sulfuric acid	98.08	1.8318	10.4	335.5	—	∞	∞	—

Note: i: insoluble; s: soluble; vs: very soluble; ∞: miscible.

【Reagent dosage】

Reagent	Dosage
Water	3 mL
Concentrated sulfuric acid	4 mL(7.3 g, 0.074 mol)
n-Butyl alcohol	3.0 mL(2.4 g, 0.033 mol)
Sodium bromide	4.0 g(0.039 mol)

Ⅳ Procedures

1. Apparatus

The reaction device contains a 25 mL round-bottom flask, the reflux apparatus and gas trap shown in Figure 4-1. For the gas trap, add 5% aqueous sodium hydroxide to the beaker with an inverted funnel just above surface. The trap absorbs the hydrogen bromide gas evolved during the reaction period.

2. Reaction mixture

Place 3 mL of water in the round-bottom flask and slowly add 4 mL of concentrated sulfuric acid with continuous swirling. Cool the mixture to room temperature and add 3.0 mL of *n*-butyl alcohol and 4.0 g of sodium bromide. Add several boiling stones to the mixture and place the flask in a heating mantle, clamp it securely, and connect it with a Allihn condenser for reflux condensation. Be sure not to merge the inverted funnel of the gas trap into aqueous solution in order to avoid inverse suction.

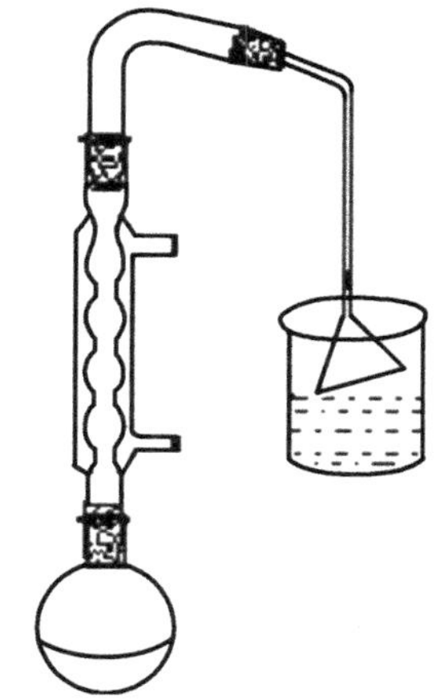

Figure 4-1　Apparatus for Preparing *n*-Butyl Bromide

3. Reflux

Heat the mixture to a gentle boil in the heating mantle, swirling frequently. The upper layer that soon separates is the alkyl bromide. Continue heating the mixture for 30 min. Then stop heating and allow the apparatus to cool before disconnecting the reflux apparatus.

4. Distillation

Remove the condenser with the gas trap and assemble a simple distillation apparatus. Mount a distilling head in the flask, and set the liebig condenser for downward distillation through a bent or vacuum adapter into a 50 mL Erlenmeyer flask. Distill *n*-butyl bromide until the distillate appears to be clear [1].

5. Extractions

Pour the distillate into a 50 mL separatory funnel and add the equivalent volume of water to wash the mixture [2]. Stopper the funnel and shake it, venting occasionally. Allow the layers to separate. The *n*-butyl bromide layer should be lower. Drain the lower layer through the stopcock into another dry separatory funnel. Add the equivalent volume of concentrated sulfuric acid to the second funnel and shake the mixture [3]. Allow the layers to separate. Drain and discard the lower sulfuric acid layer. Extract the organic layer with H_2O again, and then with saturated aqueous sodium bicarbonate. Note that the upper water phase should be separated from the lower organic phase as completely as possible in the last extraction. A detailed flow sheet for the isolation and purification is shown as follows:

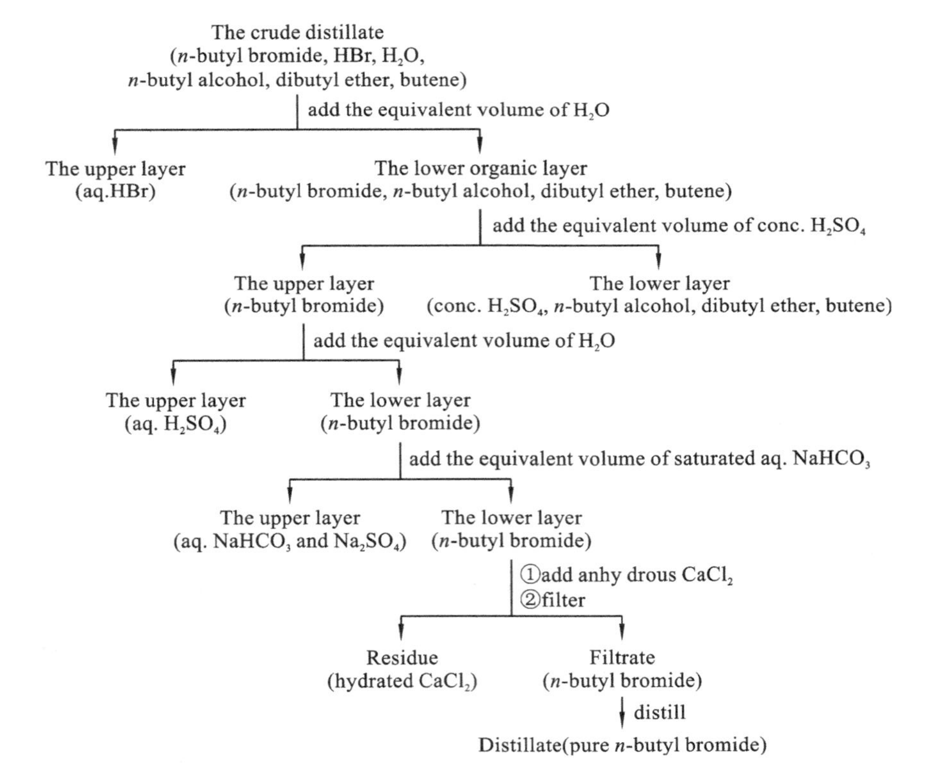

6. Drying

Drain the alkyl halide layer into a dry Erlenmeyer flask. Add about 0.5 g of anhydrous calcium chloride to dry the solution for about 20 min until the liquid is clear [4].

7. Distillation

Pour the dried liquid into a dry round-bottom flask carefully, remaining the drying agents in the Erlenmeyer flask(or filter the liquid into the flask). Add boiling stones and distill the crude *n*-butyl bromide. Collect the distillate boiling in the range of 98～102 ℃.

8. Yield determination

Weigh the product and calculate the percentage yield. 1. 8～2. 5 g of the product is obtained with the yield of 40%～55%. Determine the refractive index. Pure *n*-butyl bromide is a colorless liquid.

This experiment requires 5 h.

Ⅴ Notes and Instructions

【Notes】

[1] You can judge the ending of the distillation of *n*-butyl bromide in the following three ways:

① The distillate turns from cloudy to clear;

② The upper organic layer disappears in the reaction flask;

③ Collect a few drops of the distillate from the adapter into a water-contained beaker, shaking, and be sure that no oil droplet appears. The test indicates that there is no organic compound in the distillate and the distillation is completed.

[2] If the color of distillate is pink after washing with water, it shows that a trace of bromine is in the solution. Bromine can be removed by using saturated sodium bisulfite to extract.

$$2HBr + H_2SO_4 \longrightarrow Br_2 + SO_2\uparrow + 2H_2O$$

$$Br_2 + 3NaHSO_3 \longrightarrow 2NaBr + NaHSO_4 + 2SO_2\uparrow + H_2O$$

[3] A small quantity of unreacted *n*-butyl alcohol and byproducts, such as dibutyl ether, should be removed by using concentrated sulfuric acid to extract. Otherwise, it is difficult to get rid of them in the following distillation because of the formation of binary azeotrope of *n*-butyl alcohol and *n*-butyl bromide. The azeotrope containing 13% *n*-butyl alcohol distills at 98. 6 ℃.

[4] Anhydrous calcium chloride as a drying agent can absorb alcohols in addition of water, because alcohol can complex with calcium chloride which is insoluble in organic solvent.

【Requirements for preview】

(1) Get to know the influencing factors of the reaction rate and yield of product.

(2) Learn about the category and usage of the gas trap and drying agents (see Experiment 2).

(3) Be familiar with the detailed flow sheet for the isolation and purification of *n*-butyl bromide. Indicate how each byproduct is removed and which layer is expected to contain the product in each separation step.

【Experimental precautions】

(1) Concentrated sulfuric acid is highly corrosive and causes severe burns. Please take special care with it.

(2) A gas trap is often used to absorb harmful gas evolved during a reaction. There is only a few amount of HBr gas produced in this experiment, so water can be used as an absorbent in place of NaOH solution.

(3) Make sure to add the reactants in order. Place water, concentrated sulfuric acid, *n*-butyl alcohol and sodium bromide into the reaction flask in turn. Here, water is important for this reaction:

① to reduce the volatilization of hydrogen bromide gas;

② to prevent the generation of foam;

③ to decrease the concentration of sulfuric acid to reduce the generation of byproduct.

(4) The mixture should be swirled frequently in the period of adding the materials to the flask and reflux, otherwise the yield of the product decreases. The reasons for swirling are as follows:

① Slowly add concentrated sulfuric acid to water while swirling to prevent the solution from being overheated since it is a violent exothermic process;

② Add *n*-butyl alcohol while swirling to prevent it being carbonized by concentrated sulfuric acid;

③ Add sodium bromide while swirling to avoid forming blocks and promote the production of hydrogen bromide;

④ Swirling two phases during reflux can promote the molecules in two layers getting in touch with each other more efficient and drive the reaction to complete.

(5) The residue should be poured out of the flask immediately after distillation is complete; otherwise it turns into block and is difficult to be cleaned up.

Ⅵ Post-lab Questions

(1) Why should the separatory funnel for extraction with concentrated sulfuric acid be dry?

(2) Write out the reaction of anhydrous calcium chloride with water. Before distillation, the crude liquid should be filtered after drying with anhydrous calcium chloride, why?

(3) In this experiment, *n*-butyl bromide of the distillate usually stays in the lower layer. But it occasionally appears in the upper layer, why? How to deal with this situation?

Ⅶ Verbs

substitution reaction 取代反应;

n-butyl alcohol 正丁醇;

n-butyl bromide 正溴丁烷;

concentrated sulfuric acid 浓硫酸;

sodium bromide 溴化钠;

sodium bicarbonate 碳酸氢钠;

calcium chloride 氯化钙;

sodium bisulfite 亚硫酸氢钠

实验十八　正丁醚的制备

一、实验目的

（1）掌握醇分子间脱水制备醚的反应原理和实验方法。

（2）学习使用分水器的实验操作。

二、实验原理

正丁醇在酸催化下进行分子间脱水反应生成正丁醚。主反应如下：

$$2CH_3CH_2CH_2CH_2OH \underset{135\ ℃}{\overset{H_2SO_4}{\rightleftharpoons}} CH_3CH_2CH_2CH_2OCH_2CH_2CH_2CH_3 + H_2O$$

这是一个可逆反应，为获得较好收率，本实验主要采用分水器使生成的水迅速脱离反应区，使平衡向右移动生成产物。

反应过程中需要严格控制温度，135 ℃以上正丁醇易在酸催化下进行分子内脱水生成烯烃。副反应如下：

$$CH_3CH_2CH_2CH_2OH \xrightarrow{H_2SO_4} CH_3CH_2CH{=}CH_2 + H_2O$$

三、仪器与试剂

【仪器】三口烧瓶（25 mL），圆底烧瓶（10 mL），球形冷凝管，分水器，温度计，分液漏斗（50 mL），锥形瓶（10 mL），蒸馏头，直形冷凝管，尾接管，漏斗，折光仪。

【试剂】正丁醇，浓硫酸，5% 氢氧化钠溶液，饱和氯化钙溶液，无水氯化钙。

【物理常数】

化合物名称	相对分子质量	密度/（g·cm^{-3}）	熔点/℃	沸点/℃	折光率（n_D^{20}）	溶解度/[g·(100 mL)$^{-1}$]		
						水	乙醇	乙醚
正丁醇	74.12	0.8097	−89.5	117.7	1.3993	7.4	∞	∞
1-丁烯	56.11	0.6700	−185.3	−6.5	1.3962	i	vs	vs
正丁醚	130.22	0.7689	−95	140	1.3992	0.03	∞	∞
浓硫酸	98.08	1.8318	10.4	335.5	—	∞	∞	—

注：i 表示不溶；vs 表示易溶；∞表示混溶。

【试剂用量】正丁醇，10.3 mL(8.3 g，0.112 mol)；浓硫酸，1.7 mL(3.1 g，0.032 mol)。

四、实验步骤

1. 反应加料

在 25 mL 三口烧瓶中，加入 10.3 mL 正丁醇，然后逐步加入 1.7 mL 浓硫酸（边加边振摇）。添加两粒沸石后，把反应瓶置于电热套中，用夹子加紧，按图 4-2 所示安装仪器。三口烧瓶的一侧口插上温度计，温度计水银球应浸入液面以下；中间装上分水器，分水器的上端接回流冷凝管；另一口用塞子塞紧。分水器用水检漏后，预先在分水器内放置双蒸水至接近支管口。

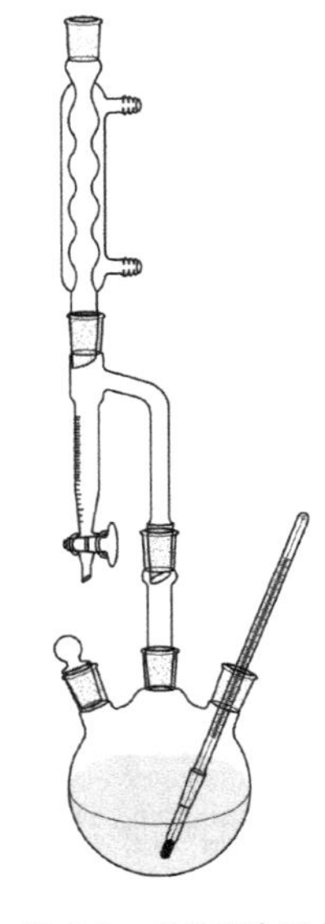

图 4-2 正丁醚制备反应装置

2. 回流

开启冷凝水,将三口烧瓶放在电热套上小火加热至微沸回流。回流初始阶段,应尽量使回流的有机物返回反应瓶中;随着反应生成的水增多,回流液里开始出现水珠,此时应从分水器中逐渐放出一部分水(总共约 1.2 mL)[1],收集于分水器内的回流液里的水珠会下降并沉于下层,上层有机相沉积至分水器支管时,即可再次返回烧瓶。这个循环持续进行,直至分水器中水层量不再增加为止。当三口烧瓶中反应液温度升至 135 ℃左右[2],此时分水器全部被水充满,即可停止反应,大约需要 45 min。若继续加热,则反应液变黑并有较多副产物烯烃生成。除去热源,将反应液冷却至室温。

3. 萃取

预先在分液漏斗中放入 15 mL 水,并将分液漏斗置于铁架台的铁圈中。拆除回流装置,将反应液通过漏斗倒入分液漏斗中(注意不要将沸石倒入其中)。塞紧瓶塞,充分振摇分液漏斗,静置分层后弃去下层液体。上层粗产物依次用 8 mL 水、5 mL 5%氢氧化钠溶液、5 mL 水和 5 mL 饱和氯化钙溶液洗涤。正丁醚分离纯化流程如下[3]:

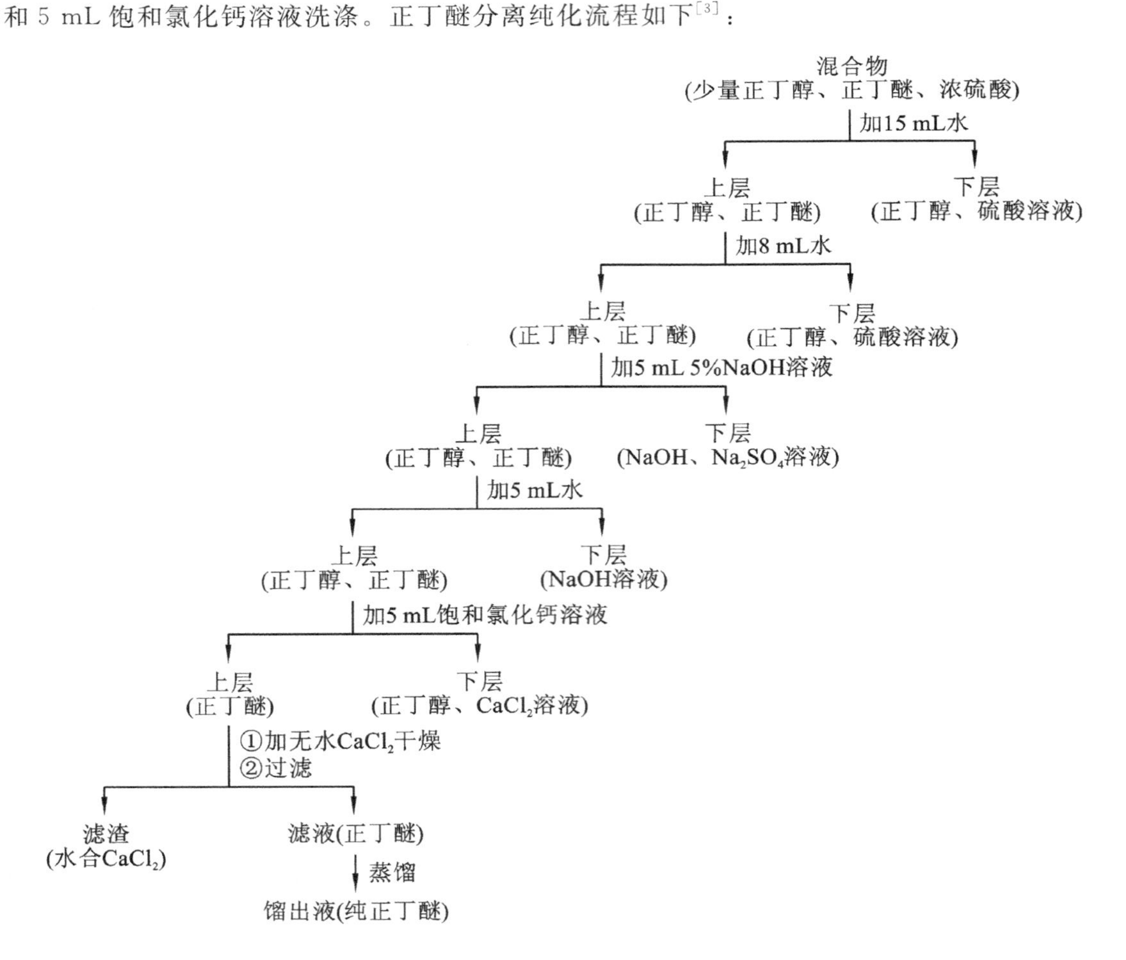

4. 干燥

将上层有机相转入干燥的锥形瓶中，用适量无水氯化钙干燥 20 min(此时，可以准备蒸馏装置)。

5. 蒸馏

将干燥后的粗产物滤入 10 mL 圆底烧瓶中，加入一粒沸石，预先称重小的锥形瓶，用来接收产物。蒸馏并收集 140～144 ℃馏分。

6. 产率测定

称重，产量约为 2.5 g(产率约为 34%)。测定产物的折光率。纯正丁醚为无色透明液体。

本实验约需 5 h。

五、注解和实验指导

【注解】

[1] 本实验根据理论计算产生水体积为 1 mL，但实际分出水层的体积略大于计算量，故最后从加满水的分水器中总共放掉约 1.2 mL 水。

[2] 制备正丁醚的最佳温度是 135～140 ℃，但此温度在开始回流时是很难达到的。因为正丁醚可与水形成共沸物(沸点 94.1 ℃，含水 33.4%)；另外，正丁醚与水及正丁醇形成三元共沸物(沸点 90.6 ℃，含水 29.9%，正丁醇 34.6%)，正丁醇也可与水形成共沸物(沸点 93.0 ℃，含水 44.5%)，故温度控制在 90～100 ℃较合适，而实际操作是在 100～105 ℃。

[3] 上层粗产物的洗涤也可采用下法进行：先每次用 8 mL 50%硫酸洗涤 2 次，再每次用 8 mL 水洗涤 2 次。因 50%硫酸可洗去粗产物中的正丁醇，但正丁醚也能微溶，所以产率略有降低。另外，由于氯化钙能与醇形成复合物，可除去醚溶液中部分未反应的醇。

【预习要求】

(1) 分析影响醇分子间脱水反应速率和产率的因素。

(2) 理解分水器的分水原理和使用方法。

(3) 熟悉液体产物的分离提纯程序。

(4) 了解干燥剂的选择和使用(见本书第二部分实验二)。

【操作注意事项】

(1) 浓硫酸有腐蚀性，易灼伤，小心使用。

(2) 投料时须充分摇动，否则硫酸局部过浓，加热后易使反应溶液变黑。

(3) 在萃取碱洗过程中，不要太剧烈地摇动分液漏斗，否则生成乳浊液，分离困难。一旦形成乳浊液，可加入少量食盐等电解质，使之分层。

六、思考题

(1) 本实验中使用分水器的目的和作用是什么？

(2) 如何得知反应已经比较完全？

(3) 反应物冷却后为什么要倒入 15 mL 水中？各步的洗涤目的何在？

Experiment 18 Synthesis of Dibutyl Ether

Ⅰ Objectives

(1) To understand the principle of intermolecular dehydration and preparation of ethers.

(2) To learn the experimental operation of the reflux apparatus with a water trap vessel.

Ⅱ Principle

Dibutyl ether can be prepared by intermolecular dehydration of *n*-butyl alcohol catalyzed with concentrated sulfuric acid.

Main reaction:

$$2CH_3CH_2CH_2CH_2OH \xrightleftharpoons[135\ ^\circ C]{conc.\ H_2SO_4} CH_3CH_2CH_2CH_2OCH_2CH_2CH_2CH_3 + H_2O$$

The dehydration reaction is reversible under heating but in favor of the formation of ether in the presence of the concentrated sulfuric acid as a catalyst. To drive the equilibrium shifting to the right, water, the byproduct at right, should be removed from the reaction mixture continuously, which can be done with a water trap vessel.

The reaction temperature should be carefully controlled. As a side reaction, an elimination reaction to produce 1-butene, will take place in case the reaction temperature is higher than 135 ℃.

Side reaction:

$$CH_3CH_2CH_2CH_2OH \xrightarrow[\triangle]{conc.\ H_2SO_4} CH_3CH_2CH{=}CH_2 + H_2O$$

Ⅲ Apparatus and Reagents

【Apparatus】 Three-neck flask (25 mL), round-bottom flask (10 mL), reflux condenser, water trap vessel, separatory funnel (50 mL), Erlenmeyer flask, simple distillation apparatus, refractometer.

【Reagents】 *n*-Butyl alcohol, concentrated sulfuric acid, 5% aqueous sodium hydroxide, saturated aqueous calcium chloride, anhydrous calcium chloride.

【Physical constants】

Compound	M_w	ρ /(g · cm^{-3})	m. p. /℃	b. p. /℃	Refractive Index(n_D^{20})	Solubility/[g · (100 mL)$^{-1}$]		
						Water	Ethanol	Diethyl ether
n-Butyl alcohol	74.12	0.8097	−89.5	117.7	1.3993	7.4	∞	∞
1-Butene	56.11	0.6700	−185.3	−6.5	1.3962	i	vs	vs
Dibutyl ether	130.22	0.7689	−95	140	1.3992	0.03	∞	∞
Concentrated sulfuric acid	98.08	1.8318	10.4	335.5	—	∞	∞	—

Note: i: insoluble; vs: very soluble; ∞: miscible.

【Reagent dosage】*n*-Butyl alcohol, 10.3 mL(8.3 g, 0.112 mol); concentrated sulfuric acid, 1.7 mL(3.1 g, 0.032 mol).

Ⅳ Procedures

1. Reaction mixture

Place 10.3 mL of *n*-butyl alcohol in a 25 mL three-neck flask and add 1.7 mL of concentrated sulfuric acid dropwise with constant swirling. Add two boiling stones and place the flask in a heating mantle, clamp it securely. Set up the apparatus according to Figure 4-2. The bulb of the thermometer must be placed below the surface. The unused third neck is corked with a stopper. The water trap vessel must be checked for leaks, and then is filled with water at the side with a stopcock in advance.

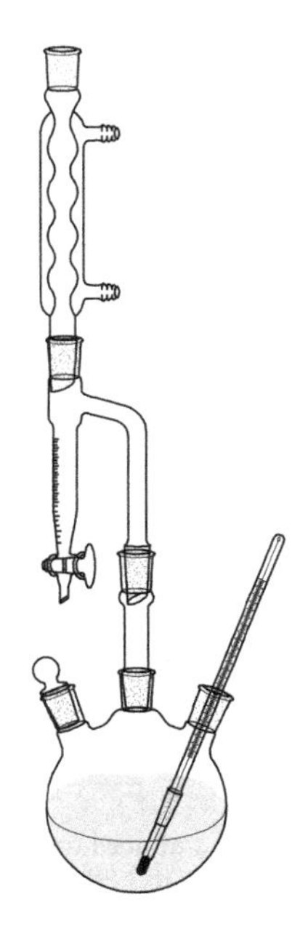

Figure 4-2 Apparatus for the Preparation of Dibutyl Ether

2. Reflux

Start water circulating in the condenser and heat the mixture slowly to boil by a heating mantle. At the beginning of reflux, the condensate with unreacted organic reactants should flow back to the flask for further heating instead of being trapped in the vessel. As the reaction occurs faster and more and more water has been produced, the condensate must be trapped in order to collect the water produced. So you should let the water previously placed in the vessel drain out from the stopcock gradually(1.2 mL in all) when you have observed the formation of water droplets in the condensate [1]. When the condensate is collected in the water trap vessel, the water droplets it contains will drop to the lower water phase and the organic layer is above. When the surface of upper layer is higher than the sidearm, the organic liquid will overflow back into the reaction flask. The cycle is repeated continuously, until no more water forms. After 45 min, the temperature of the reaction mixture increases to approximately 135 ℃[2]. Now the water trap vessel is filled with water again, that shows the reaction is nearly completed. Then stop heating and wait for the mixture to cool down before you disconnect the reflux apparatus.

3. Extractions

Place a separatory funnel in an iron ring attached to a ring stand and add 15 mL of water into it. Disassemble the apparatus and pour the reaction mixture into the separatory funnel with care. Also be careful to avoid transferring the boiling stones, or you will need to remove it after the transfer. Stopper the funnel and shake it, venting occasionally. Allow the phases to separate and then discard the lower aqueous layer through the stopcock into a beaker. Next, extract the upper organic layer with 8 mL of water, 5 mL of 5% aqueous sodium hydroxide, 5 mL of water and 5 mL of saturated aqueous calcium chloride in turn. A detailed flow sheet for the isolation and purification of n-butyl ether is as follows[3]:

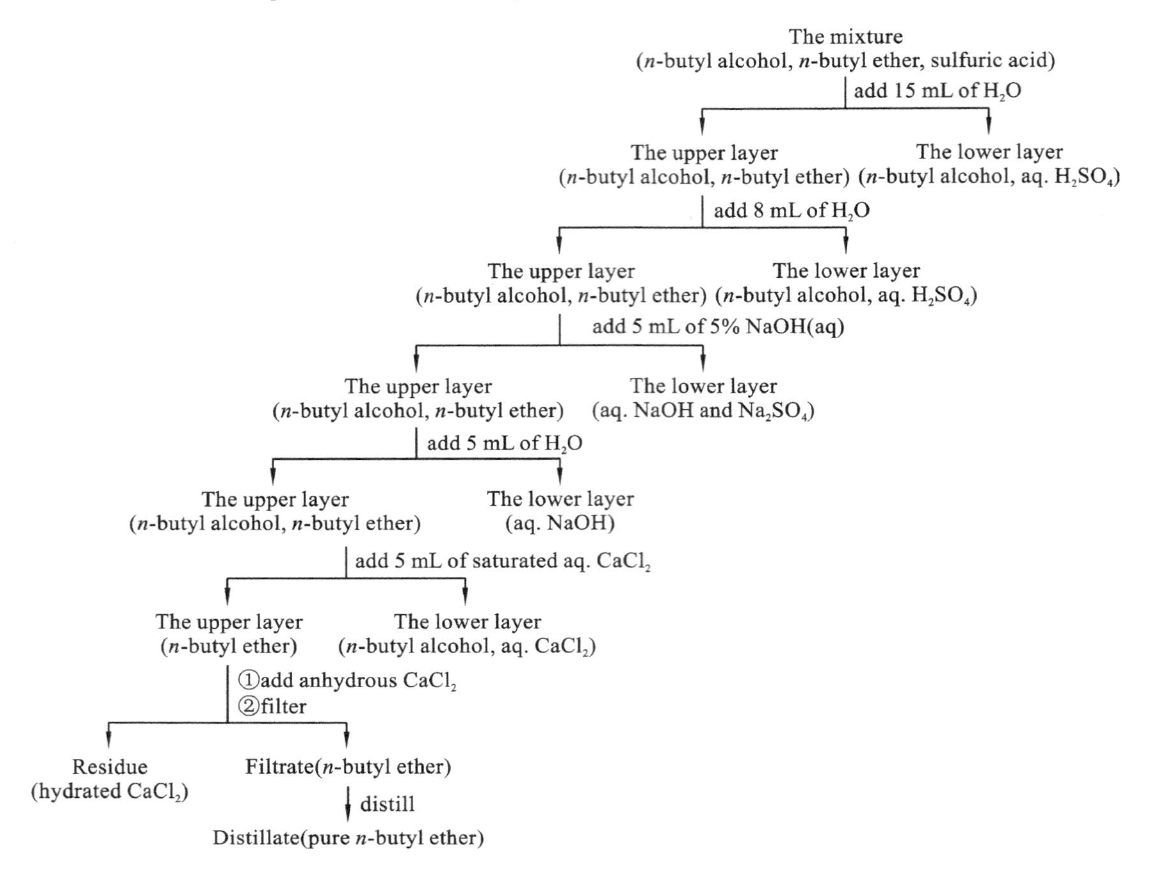

4. Drying

Transfer the crude ether to a clean, dry Erlenmeyer flask. Add 1 g of anhydrous calcium chloride. Stand the flask without disturbance for 20 min until the liquid turns clear.

5. Distillation

Filter the dried liquid into a dry 10 mL round-bottom flask. Add a boiling stone, continue simple distillation and collect the distillate boiling in the range of 140~144 ℃. Tare a dry Erlenmeyer flask to collect the product.

6. Yield determination

Weigh the product and calculate the percentage yield. About 2.5 g of the product is obtained with the yield of 34%. Pure dibutyl ether is a colorless liquid. Determine the refractive index.

This experiment requires 5 h.

Ⅴ Notes and Instructions

【Notes】

[1] In this experiment, 1 mL of water should be produced theoretically. In fact, the

volume of water collected is slightly greater than the calculated amount. So you should drain 1.2 mL of water through the stopcock of water trap vessel in all.

[2] The intermolecular dehydration reaction of *n*-butyl alcohol catalyzed by concentrated sulfuric acid takes place in the range of 135～140 ℃. The mixture under reflux can form the azeotrope initially, which makes the reaction temperature is difficult to reach 135 ℃. Dibutyl ether-water azeotrope(33.4% water) distills at 94.1 ℃. Dibutyl ether-water-*n*-butyl alcohol azeotrope (29.9% water, 34.6% *n*-butyl alcohol) distills at 90.6 ℃. *n*-Butyl alcohol-water azeotrope (44.5% water) distills at 93 ℃. So the temperature of reaction mixture is controlled in the range of 90～100 ℃. However, the operation temperature is 100～105 ℃.

[3] As an alternative, the upper organic layer may be extracted in the following way: twice with 8 mL of 50% aqueous sulfuric acid and twice with 8 mL of water in turn. *n*-Butyl alcohol can dissolve in 50% aqueous sulfuric acid, so it can be removed from the crude dibutyl ether. But dibutyl ether is slightly soluble in 50% aqueous sulfuric acid, which may decrease the yield a little. In addition, calcium chloride can get rid of remaining alcohol in ether because it forms complex with alcohol.

【Requirements for preview】

(1) Review the reaction conditions for intermolecular dehydration of alcohols to prepare ethers. Analyze the influencing factors that affect the reaction rate and yield of alcohol intermolecular dehydration.

(2) Understand the working principle of a water trap vessel and its usage.

(3) Be familiar with the isolation and purification procedure for a liquid product.

(4) Learn about the types and usage of drying agents (see Experiment 2).

【Experimental precautions】

(1) Concentrated sulfuric acid is highly corrosive and causes severe burns. Please take special care with it.

(2) When concentrated sulfuric acid is added to the reaction mixture, mix immediately and fully(swirl). Otherwise, *n*-butanol can be easily oxidized by concentrated sulfuric acid and the solution will turn black under heating.

(3) Don't shake the separatory funnel vigorously during the extraction by base; the solution will form emulsion when you wash the organic layer with base. It is difficult to separate the emulsion. However, sodium chloride can destroy the emulsion and allow the phases to separate.

Ⅵ Post-lab Questions

(1) What's the purpose of using a water trap vessel in the preparation of dibutyl ether?

(2) How do you know the synthesis of dibutyl ether is nearly completed?

(3) Why should the reaction mixture be poured into the separatory funnel containing 15 mL of water? What's the purpose of each separation step?

Ⅶ Verbs

intermolecular dehydration 分子间脱水;
water trap vessel 分水器;
n-butyl alcohol 正丁醇;
dibutyl ether 正丁醚;
concentrated sulfuric acid 浓硫酸;
sodium hydroxide 氢氧化钠;
calcium chloride 氯化钙

实验十九 乙酸正丁酯的制备

一、实验目的

(1) 学习通过有机酸和醇的酯化反应制备酯类化合物的原理和方法。

(2) 掌握液体产物的分离提纯方法,巩固萃取、液体洗涤及干燥、蒸馏等基本操作。

二、实验原理

乙酸正丁酯是通过乙酸和正丁醇的直接酯化反应制备得到的。主反应式如下:

$$CH_3COOH + n\text{-}C_4H_9OH \xrightleftharpoons[\triangle]{\text{浓 } H_2SO_4} CH_3COOC_4H_9\text{-}n + H_2O$$

酯化反应是可逆反应,为了使反应向有利于生成酯的方向移动,通常采用过量的羧酸或醇,或者除去反应中生成的酯或水,或者两者同时采用。本实验是利用冰乙酸过量使反应向正方向进行,从而提高反应产率。

副反应式如下:

$$2CH_3CH_2CH_2CH_2OH \xrightleftharpoons[\triangle]{\text{浓 } H_2SO_4} CH_3CH_2CH_2CH_2OCH_2CH_2CH_2CH_3 + H_2O$$

$$CH_3CH_2CH_2CH_2OH \xrightarrow[\triangle]{\text{浓 } H_2SO_4} CH_3CH_2CH{=}CH_2 + H_2O$$

反应副产物主要有丁醚和丁烯。在分离纯化过程中,过量的乙酸和残余的正丁醇可以分别通过碳酸氢钠和水除去。粗产物通过无水硫酸镁干燥后,通过蒸馏进行纯化。

三、仪器与试剂

【仪器】圆底烧瓶(25 mL、10 mL),回流冷凝管,分液漏斗(50 mL),锥形瓶,普通蒸馏装置,折光仪。

【试剂】正丁醇,冰乙酸,浓硫酸,5% 碳酸氢钠溶液,饱和氯化钠溶液,无水硫酸镁。

【物理常数】

名　　称	相对分子质量	密度/($g \cdot cm^{-3}$)	熔点/℃	沸点/℃	折光率(n_D^{20})	溶解度/[$g \cdot (100\ mL)^{-1}$]		
						水	乙醇	乙醚
正丁醇	74.12	0.8097	−89.5	117.7	1.3993	7.4	∞	∞
冰乙酸	60.05	1.0492	16.7	118	1.3718	∞	∞	∞
乙酸正丁酯	116.16	0.8813	−78	126	1.3941	0.43	∞	∞

续表

名　　称	相对分子质量	密度/(g·cm^{-3})	熔点/℃	沸点/℃	折光率(n_D^{20})	溶解度/[g·(100 mL)$^{-1}$]		
						水	乙醇	乙醚
1-丁烯	56.11	0.6700	−185.3	−6.5	1.3962	i	vs	vs
正丁醚	130.22	0.7689	−95	140	1.3992	0.03	∞	∞

注：i 表示不溶；vs 表示易溶；∞表示混溶。

【试剂用量】

试　　剂	用　　量
正丁醇	3.0 mL(2.4 g,0.033 mol)
冰乙酸	4.3 mL(4.5 g,0.075 mol)
浓硫酸	0.9 mL(1.6 g,0.017 mol)

四、实验步骤

1. 安装实验装置

安装回流装置(见图 4-3)，包括圆底烧瓶和回流冷凝管。[1]采用电热套进行加热。

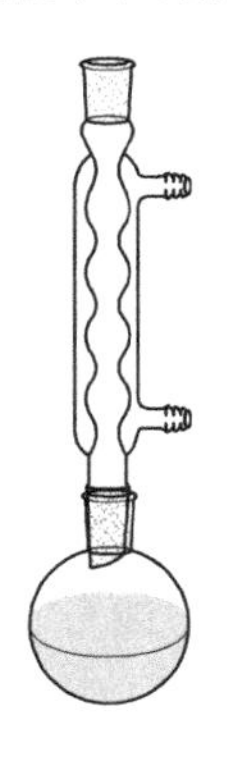

图 4-3　乙酸正丁酯合成装置

2. 加入反应物

在 25 mL 圆底烧瓶中加入 3.0 mL 正丁醇、4.3 mL 冰乙酸，边摇动边慢慢滴加 0.9 mL 浓硫酸，塞上空心塞，混匀后加入两粒沸石。

3. 回流

将圆底烧瓶置于电热套上，用夹子夹好，装上回流冷凝管。开启冷凝水，反应物在电热套上加热回流 1 h。

4. 萃取

将反应物冷却至室温，小心移入分液漏斗中，用 10 mL 冷水洗涤圆底烧瓶，并将洗液合并于分液漏斗中。振摇后静置，分出下层水层，有机相用 5 mL 5%碳酸氢钠溶液洗涤[2]，至水溶液呈弱碱性。然后用 5 mL 饱和氯化钠溶液洗涤[3]，分去水层。产物分离纯化流程如下：

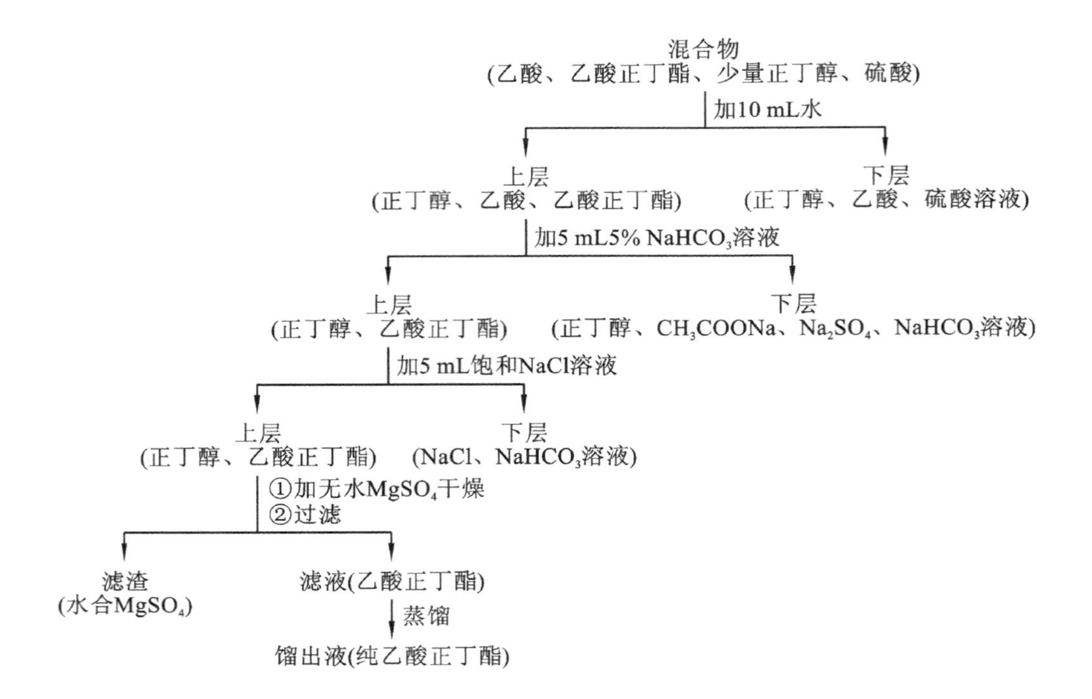

5. 干燥

将酯层转移至一个干燥的小锥形瓶中,用约 0.5 g 无水硫酸镁干燥[4]。轻轻摇动,至有机层透明,干燥 10～15 min。

6. 蒸馏

将粗产物滤入 10 mL 圆底烧瓶中,加入一粒沸石,安装好蒸馏装置,用一个已称重的锥形瓶接收产品。加热蒸馏,收集 122～126 ℃的馏分。

7. 测定产率

称重,产量约为 2.0 g(产率约为 53%)。测定产物的折光率。纯乙酸正丁酯是无色液体。本实验约需 5 h。

五、注解和实验指导

【注解】

[1] 本实验也可采用分水器回流装置进行反应,通过不断去除反应中生成的水促使平衡向生成酯的方向移动,以提高产率。参见本书实验十八。

[2] 用碳酸氢钠溶液洗涤时,因为有大量的 CO_2 气体放出,所以开始时不要塞住分液漏斗,敞开振摇漏斗至无明显的气泡产生后再塞住振摇,洗涤时应注意及时放气。

[3] 饱和氯化钠溶液可降低酯在水中的溶解度,还可以防止乳化,有利于分层,便于分离。

[4] 不能用无水氯化钙干燥乙酸正丁酯,因氯化钙能与酯形成配合物。

【预习要求】

(1) 学习酯化反应的机理。

(2) 了解影响酯化反应速率和产率的因素。

(3) 熟悉粗产物乙酸正丁酯分离纯化流程，确定每一步分离纯化的目的和产物所在的液层。

(4) 了解干燥剂的种类和使用方法(见实验二)。

【操作注意事项】

(1) 酯化反应所用仪器和试剂必须无水，包括量取正丁醇和冰乙酸的量筒也要干燥。

(2) 浓硫酸和冰乙酸都有腐蚀性，要小心取用。一旦接触，要用大量水连续冲洗患处。

(3) 滴加浓硫酸时，应边加边摇，必要时可用冷水冷却，以免局部炭化。反应物与浓硫酸一定要混合均匀，若不均匀，加热时会使有机物炭化，溶液发黑。

六、思考题

(1) 用无水硫酸镁干燥粗乙酸正丁酯时，如何掌握干燥剂的用量？

(2) 在乙酸正丁酯的精制过程中，如果最后蒸馏时前馏分多，其原因是什么？

Experiment 19　Synthesis of *n*-Butyl Acetate

Ⅰ Objectives

(1) To understand the principle of esterification reactions of organic acid with alcohol.

(2) To learn the isolation and purification methods of synthetic liquid products by extraction, washing, drying and distillation procedures.

Ⅱ Principle

n-Butyl acetate can be prepared by the direct esterification of acetic acid with *n*-butyl alcohol.

Main reaction:

$$CH_3COOH + n\text{-}C_4H_9OH \xrightleftharpoons[\triangle]{\text{conc. } H_2SO_4} CH_3COOC_4H_9\text{-}n + H_2O$$

Esterification is usually reversible. And the reaction equilibrium does not favor the formation of the ester. In order to obtain a higher yield of ester, the reaction must be driven to shift to the right. By using an excess amount of starting materials (usually the one can be easily removed from reaction mixture) or by removing water continuously from the reaction system, the equilibrium will be favored to the direction of ester formation. In this experiment, glacial acetic acid is excessively added to drive the equilibrium to the right which favors the transformation of *n*-butyl alcohol and acetic acid to *n*-butyl acetate fully.

Side reactions:

$$2CH_3CH_2CH_2CH_2OH \xrightleftharpoons[\triangle]{\text{conc. } H_2SO_4} CH_3CH_2CH_2CH_2OCH_2CH_2CH_2CH_3 + H_2O$$

$$CH_3CH_2CH_2CH_2OH \xrightarrow[\triangle]{\text{conc. } H_2SO_4} CH_3CH_2CH{=}CH_2 + H_2O$$

Dibutyl ether and 1-butene are the possible byproducts. In the isolation procedure, the excess acetic acid and the remaining *n*-butyl alcohol are removed by extraction with sodium bicarbonate and water, respectively. After drying the crude product with anhydrous magnesium sulfate, the ester is purified by distillation.

Ⅲ Apparatus and Reagents

【Apparatus】 Round-bottom flask (25 mL, 10 mL), Allihn condenser, separatory funnel (50 mL), Erlenmeyer flask, graduated dropper, simple distillation apparatus, refractometer.

【Reagents】 *n*-Butyl alcohol, glacial acetic acid, concentrated sulfuric acid, 5% aqueous sodium bicarbonate, saturated aqueous sodium chloride, anhydrous magnesium sulfate.

【Physical constants】

Compound	M_w	ρ /(g · cm^{-3})	m. p. /℃	b. p. /℃	Refractive Index(n_D^{20})	Solubility/[g · (100 mL)$^{-1}$]		
						Water	Ethanol	Diethyl ether
n-Butyl alcohol	74.12	0.8097	−89.5	117.7	1.3993	7.4	∞	∞
Glacial acetic acid	60.05	1.0492	16.7	118	1.3718	∞	∞	∞
n-Butyl acetate	116.16	0.8813	−78	126	1.3941	0.43	∞	∞
1-Butene	56.11	0.6700	−185.3	−6.5	1.3962	i	vs	vs
Dibutyl ether	130.22	0.7689	−95	140	1.3992	0.03	∞	∞

Note: i: insoluble; vs: very soluble; ∞: miscible.

【Reagent dosage】

Reagent	Dosage
n-Butyl alcohol	3.0 mL(2.4 g, 0.033 mol)
Glacial acetic acid	4.3 mL(4.5 g, 0.075 mol)
Concentrated sulfuric acid	0.9 mL(1.6 g, 0.017 mol)

Ⅳ Procedures

1. Apparatus

Assemble a reflux apparatus shown in Figure 4-3, including a round-bottom flask and a reflux condenser [1]. All the glassware must be dry.

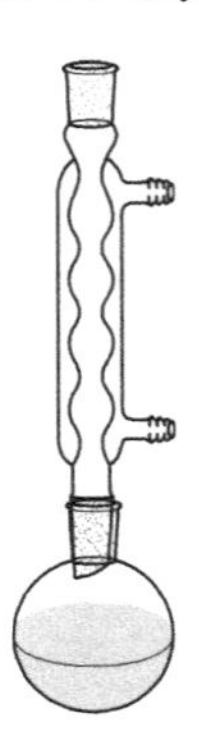

Figure 4-3 Apparatus for Preparing *n*-Butyl Acetate

2. Reaction mixture

Place 3.0 mL of *n*-butyl alcohol in a dry 25 mL round-bottom flask and add 4.3 mL of glacial acetic acid to the flask. Using a graduated dropper, add 0.9 mL of concentrated sulfuric acid under constant swirling. Place two boiling stones in the flask.

3. Reflux

Place the flask in a heating mantle, clamp it securely, and fit it with a Allihn condenser. Start water circulating in the condenser and bring the mixture to the boil. Continue heating under reflux for 1 h.

4. Extractions

Stop heating and allow the apparatus to cool down until no more condensate appears. Disassemble the condenser and pour the liquid into a separatory funnel. Use 10 mL of water to wash the reaction flask and put the residue together with the reaction mixture in the separatory funnel. Stopper the funnel and shake it, venting occasionally. Allow the phases to separate and discard the lower aqueous layer through the stopcock into a beaker. Next, extract the organic layer with 5 mL of 5% aqueous sodium bicarbonate [2] and 5 mL of saturated aqueous sodium chloride [3] in turn. A detailed flow sheet for the isolation and purification of *n*-butyl acetate is as follows:

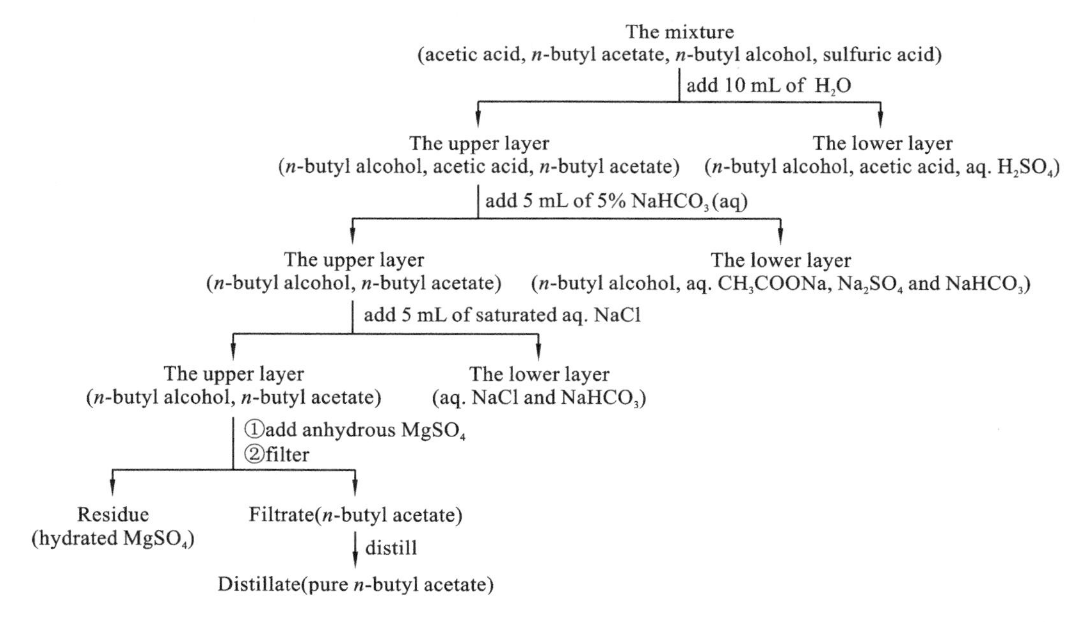

5. Drying

Transfer the crude ester to a clean, dry Erlenmeyer flask. Add 0.5 g of anhydrous magnesium sulfate [4]. Cork the mixture, allow it to stand for 10～15 min until the liquid is clear.

6. Distillation

Filter or carefully pour the dried liquid(without the drying agent!) into a dry 10 mL round-bottom flask. Add a boiling stone, assemble a distillation apparatus. Preweigh(tare) a 10 mL Erlenmeyer flask to collect the product. Continue distillation and collect the distillate boiling in the range of 122～126 ℃.

7. Yield determination

Weigh the product and calculate the percentage yield. About 2.0 g of the product is obtained with the yield of 53%. Determine the refractive index. Pure *n*-butyl acetate is a colorless liquid.

The experiment requires about 5 h.

Ⅴ Notes and Instructions

【Notes】

[1] This experiment also can be carried out with a water trap reflux apparatus, just as described in the preparation of dibutyl ether(see Experiment 18). A water trap vessel can be used to remove water, the byproduct of the reaction, from the reaction system continuously to speed up the esterification.

[2] Extract the organic layer with aqueous sodium bicarbonate. Be careful to unstopper the funnel at the beginning because acetic acid or sulfuric acid reacts with base to release carbon dioxide gas. Then stopper the funnel and shake it, venting occasionally after carbon dioxide gas disappears obviously.

[3] Saturated aqueous sodium chloride can decrease the water solubility of ester, prevent the emulsion, and favor the layers to separate.

[4] Anhydrous calcium chloride can not be used to dry the *n*-butyl acetate because it can form complex with ester.

【Requirements for preview】

(1) Get to learn the mechanism of esterification reaction of alcohol with acid.

(2) Get to know the influencing factors which affect the esterification rate and yield.

(3) Be familiar with the detailed flow sheet for the isolation and purification of *n*-butyl acetate. Indicate what byproduct is removed and which layer is expected to contain the product in each separation step.

(4) Learn about the types and usage of drying agents (see Experiment 2).

【Experimental precautions】

(1) The instrument should be dried for the esterification and the starting *n*-butyl alcohol and acetic acid should be anhydrous.

(2) Be very careful to transfer sulfuric acid and acetic acid. They are highly corrosive and keep them from directive contact with your skin.

(3) When the concentrated sulfuric acid is added to the reaction mixture, mix immediately and fully(swirl). Otherwise, the mixture may be easily oxidized and carbonized

by concentrated sulfuric acid and turn black while heating.

Ⅵ Post-lab Questions

(1) When the crude *n*-butyl acetate is dried, how to judge if the amount of anhydrous magnesium sulfate is enough or not?

(2) What compounds are there in the front distillates(those been distilled out before the collection temperature)? What are the reasons for having much front distillates when the crude *n*-butyl acetate is distilled?

Ⅶ Verbs

esterification 酯化；
n-butyl alcohol 正丁醇；
glacial acetic acid 冰乙酸；
concentrated sulfuric acid 浓硫酸；
sodium bicarbonate 碳酸氢钠；
magnesium sulfate 硫酸镁

实验二十　阿司匹林的制备

一、实验目的

(1) 掌握酰化反应的原理和阿司匹林的制备方法。
(2) 掌握有机合成中固体产物的分离提纯方法，巩固重结晶、抽滤等基本操作。
(3) 了解通过熔点测定和化学试剂检测产品纯度的方法。

二、实验原理

阿司匹林(aspirin)即乙酰水杨酸，是19世纪末成功开发并沿用至今的解热镇痛药物。其最常用的制备方法是将水杨酸与乙酸酐作用，通过乙酰化反应，使水杨酸分子中酚羟基成酯，生成乙酰水杨酸。为了打破水杨酸分子内氢键，通常加入少量的浓硫酸作为催化剂以加速反应的进行。反应式如下：

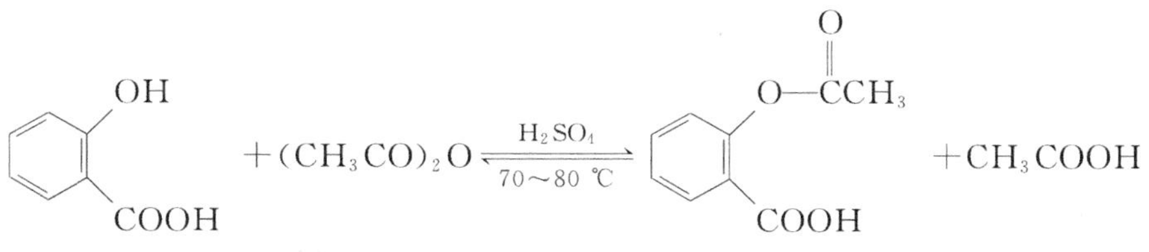

$$o\text{-}HOC_6H_4COOH + (CH_3CO)_2O \xrightleftharpoons[70\sim80\ ^\circ C]{H_2SO_4} o\text{-}CH_3COOC_6H_4COOH + CH_3COOH$$

副反应主要是酚羟基和羧基之间的脱水反应，当温度高于90 ℃时，分子间脱水加剧从而生成不溶于水的聚合物。副反应式如下：

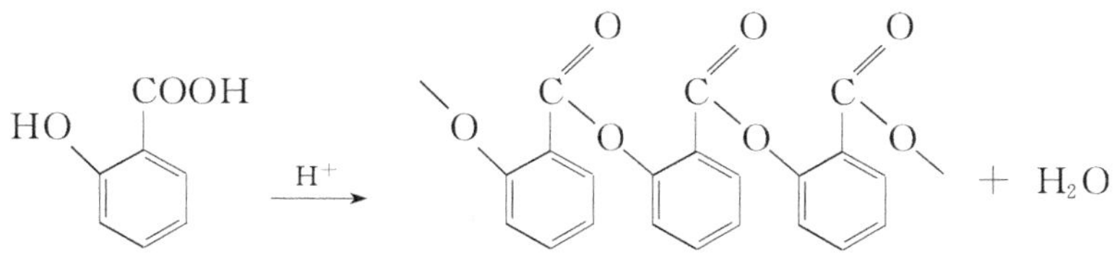

$$o\text{-}HOC_6H_4COOH \xrightarrow{H^+} \text{聚合物（酚羟基与羧基间脱水形成的聚酯）} + H_2O$$

反应生成的粗制乙酰水杨酸含有未反应完的水杨酸以及一些相对分子质量较大的副产物，可采用醇水混合溶剂进行重结晶提纯。或者先在碱溶液中溶解酸，过滤除去不溶物后再酸化,使乙酰水杨酸重结晶析出。由于很少量的酚羟基就能与 Fe^{3+} 在水溶液中形成紫红色的配合物,所以用 $FeCl_3$ 溶液检验乙酰水杨酸是否提纯是实验室最简单、常用的方法之一,有时也通过测定熔点来检测。

三、仪器与试剂

【仪器】锥形瓶,烧杯,蒸发皿,布氏漏斗,抽滤瓶,水泵,试管,熔点管。

【试剂】水杨酸,乙酸酐,浓硫酸,浓盐酸,饱和碳酸氢钠溶液,95%乙醇,1% $FeCl_3$ 溶液。

【物理常数】

化　合　物	相对分子质量	密度/(g·cm^{-3})	熔点/℃	沸点/℃	溶解度/[g·(100 mL)$^{-1}$]		
					水	乙醇	乙醚
乙酰水杨酸(阿司匹林)	180.16	1.35	136	—	0.33	20	10
水杨酸	138.12	1.44	157～159	—	0.18	37.0	33.3
浓硫酸	98.08	1.8318	10.4	335.5	∞	∞	0.98
乙酸酐	102.09	1.080	−73	139	反应	∞	∞

【试剂用量】

试　剂	用量(方法一)	用量(方法二)
水杨酸	2.0 g(0.014 mol)	6.3 g(0.045 mol)
乙酸酐	5 mL(0.05 mol)	12 mL(0.12 mol)
浓硫酸	4 滴	8～10 滴
饱和碳酸氢钠溶液	30 mL	—
95%乙醇	—	10 mL(0.21 mol)

四、实验步骤

方　法　一

1. 粗制乙酰水杨酸

蒸发皿(或水浴锅)中装适量水,加热至 80～90 ℃,作为反应水浴装置[1]。在 25 mL 干燥的锥形瓶中加入 2.0 g(0.014 moL)水杨酸和 5 mL(0.05 moL)乙酸酐,然后加入 4 滴浓硫酸,充分振摇,放置于水浴中加热溶解后继续振摇 8～10 min,然后将混合物倒入烧杯中冷却直到析出结晶。再加 36 mL 水,用冰水冷却使结晶完全析出,减压过滤,用少量(约 5 mL)蒸馏水洗涤固体两次。

2. 纯化产品

将过滤得到的粗品置于 50 mL 烧杯中,加入 30 mL 饱和碳酸氢钠溶液,充分搅拌后减压抽滤除去不溶物。滤液用浓盐酸酸化后,有晶体析出,用冰水冷却使结晶析出完全。结晶经减压抽滤、水洗、干燥后,即得较纯乙酰水杨酸,称重,计算产率。

3. 产品纯度的检测

(1) 产品干燥后研细,用毛细管法或熔点仪测定样品熔点(参见实验十三)。纯净的乙酰水杨酸固体的熔点为 135～136 ℃[2]。

(2) 取 3 支试管并编号，1 号试管中放入少量水杨酸，2 号试管中放入少量重结晶纯化后的产品，3 号试管中放入少量乙酰水杨酸标准品。分别加入蒸馏水使样品完全溶解，然后分别滴入 2～3 滴 1% $FeCl_3$溶液，观察现象并记录。

方　法　二

1. 粗制乙酰水杨酸

在 50 mL 干燥的锥形瓶中加入 6.3 g(0.045 moL)水杨酸和 12 mL(0.12 moL)乙酸酐，然后加入 8～10 滴浓硫酸，充分振摇，80～90 ℃水浴加热，固体溶解后继续振摇 8～10 min。稍微冷却后，将产物迅速倒入盛有 50 mL 蒸馏水的烧杯中，并用冷水冷却，析出晶体，如晶体难以析出，可以用玻璃棒摩擦瓶壁促使结晶形成，直至白色结晶完全析出。抽滤，并用少量(约 5 mL)蒸馏水洗涤滤饼两次，抽干，即得粗制的乙酰水杨酸。

2. 重结晶纯化产品

将粗制的乙酰水杨酸放入干燥的 100 mL 烧瓶或磨口锥形瓶中，加入 8～10 mL 乙醇，安上冷凝管加热回流，溶解后停止加热，再加入 50 mL 温水(50 ℃左右)，自然冷却到室温，有白色晶体析出后进一步冰浴冷却，使结晶析出完全。抽滤，用少量冰水洗涤滤饼，抽干，即得纯化的乙酰水杨酸。称重，计算产率。

3. 产品纯度的检测

同方法一。

五、注解和实验指导

【注解】

[1] 本实验中要注意控制好反应温度(水温低于 90 ℃)，否则将增加反应体系中羟基和羧基之间的脱水缩合，加快副产物的生成，如水杨酰水杨酸、乙酰水杨酰水杨酸、乙酰水杨酸酐以及一些聚合物等。

[2] 乙酰水杨酸在高温下易分解，分解温度为 126～135 ℃，因此测定熔点时应先将油浴加热到 120 ℃，再放入熔点管开始测定，避免样品因加热时间过长而分解。

【预习要求】

(1) 学习酰化反应，了解常用的酰化试剂。

(2) 了解阿司匹林的化学结构及其他制备方法，并与本实验中描述的方法对比。

(3) 熟悉重结晶和减压过滤等基本操作。

【操作注意事项】

(1) 实验中用到浓硫酸，应注意小心操作。

(2) 乙酸酐有挥发性，对眼睛和呼吸道有刺激，实验过程中应打开通风设备，在通风橱中取用，避免吸入。

(3) 反应用的锥形瓶以及取用试剂的量筒必须高度干燥，否则极易引起乙酸酐的水解而使实验失败。

六、思考题

(1) 在浓硫酸存在下，水杨酸与乙醇作用将得到什么产物？写出反应式。

(2) 乙醇-水体系重结晶提纯的原理是什么?

Experiment 20　Synthesis of Aspirin

Ⅰ Objectives

(1) To learn the method of preparing aspirin by acetylation reaction.

(2) To master the techniques of purification for solid products; be familiar with the operation of vacuum filtration and recrystallization.

(3) To learn the identification of purity of the solid product by determination of melting point or some specific chemical reaction.

Ⅱ Principle

Aspirin, also known as acetylsalicylic acid, is most widely sold over-the-counter drug. It has the ability to reduce fever (an antipyretic), to reduce pain (an analgesic), and to reduce swelling, soreness and redness (an anti-inflammatory agent). A useful synthesis of acetylsalicylic acid was developed in 1893, patented in 1899, marked under the trade name of "aspirin" by the Bayer Company in Germany. Despite of its side effects, aspirin remains the safest, cheapest and most effective nonprescription drug. It is made commercially, employing the same synthesis used here. In this experiment, you will prepare aspirin by the reaction of salicylic acid with acetic anhydride, using concentrated sulfuric acid as a catalyst.

Main reaction:

$$\text{C}_6\text{H}_4(\text{OH})(\text{COOH}) + (\text{CH}_3\text{CO})_2\text{O} \xrightleftharpoons[70\sim80\ ^\circ\text{C}]{\text{H}_2\text{SO}_4} \text{C}_6\text{H}_4(\text{O—}\overset{\text{O}}{\overset{\|}{\text{C}}}\text{CH}_3)(\text{COOH}) + \text{CH}_3\text{COOH}$$

At higher temperature, the crossed dehydration between the —OH groups and the —COOH groups may occur more frequently and give some polymer byproducts.

Side reaction:

$$\text{HO—C}_6\text{H}_4\text{—COOH} \xrightarrow{\text{H}^+} \text{—O—C}_6\text{H}_4\text{—C(=O)—O—C}_6\text{H}_4\text{—C(=O)—O—C}_6\text{H}_4\text{—C(=O)—O—} + \text{H}_2\text{O}$$

The crude acetylsalicylic acid needs to be further purified by removing the unreacted salicylic acid or byproducts and followed by recrystallization. The purity of final product can be detected by reacting with dilute Fe^{3+} or determining the melting point.

Ⅲ Apparatus and Reagents

【Apparatus】Erlenmeyer flask, beaker, Büchner funnel, filter flask, water aspirator, capillary tube, test tube, water bath.

【Reagents】 Salicylic acid, acetic anhydride, concentrated sulfuric acid, 95% ethanol, saturated sodium bicarbonate, concentrated hydrochloric acid, 1% $FeCl_3$.

【Physical constants】

Compound	M_w	ρ /(g·cm^{-3})	m. p. /℃	b. p /℃	Solubility/[g·(100 mL)$^{-1}$]		
					Water	Ethanol	Diethyl ether
Acetyl salicylic acid(aspirin)	180.16	1.35	136	—	0.33	20	10
Salicylic acid	138.12	1.44	157～159	—	0.18	37.0	33.3
Sulfuric acid	98.08	1.8318	10.4	335.5	Misible	Misible	0.98
Acetic anhydride	102.09	1.080	−73	139	React	Misible	Misible

【Reagents dosage】

Reagent	Dosage(Scheme 1)	Dosage(Scheme 2)
Salicylic acid	2.0 g(0.014 mol)	6.3 g(0.045 mol)
Acetic anhydride	5 mL(0.05 mol)	12 mL(0.12 mol)
Concentrated sulfuric acid	4 drops	8～10 drops
Saturated sodium bicarbonate	30 mL	—
95% Ethanol	—	10 mL(0.21 mol)

Ⅳ Procedures

Scheme 1

1. Preparation of aspirin

Prepare a water bath at 80～90 ℃ for use[1].

Weigh 2.0 g(0.014 mol) of salicylic acid and place it in a dry 50 mL Erlenmeyer flask[2]. Use this quantity of salicylic acid to calculate the theoretical or expected yield of aspirin. Carefully add 5 mL (0.05 mol) of acetic anhydride to the flask and add 4 drops of concentrated sulfuric acid with constant swirling.

Place the flask in the hot water bath for several minutes to dissolve solid material and wait for further 8～10 min to complete the reaction, and then pour the solution into a 100 mL beaker containing 36 mL of water and rinse the flask with water. Swirl to aid hydrolysis of excess acetic anhydride and then cool thoroughly in ice, scratch the inwall of the beaker with a stirring rod to induce crystallization, and collect the crystalline solid by vacuum filtration.

2. Purification of aspirin

Transfer your crude products to a 50 mL beaker and add 30 mL of saturated sodium bicarbonate solution. Swirl the solution to allow a complete neutralization between acids and sodium bicarbonate. After then, remove the insoluble substances by vacuum filtration. Pour all the filtrate to a clean beaker; slowly add concentrated hydrochloric acid into the solution while swirling the beaker until some crystalline solid comes out. Then place the beaker into an ice bath for complete crystallization. Collect the crystals by vacuum filtration. Dry the crystals under an infrared lamp. Weigh your product and calculate the yield percentage.

3. Analysis of purity

1) Detecting by Fe^{3+}

Phenols form a colored complex with the ferric ion. The purple color indicates the presence of a phenol group. Note the aspirin no longer has the phenol group. Thus a pure sample of aspirin will not give a purple color when treated with 1% $FeCl_3$.

Label three test tubes: place a few crystals of salicylic acid into test tube No. 1, a few crystals of your product into test tube No. 2, and a small amount of commercial aspirin into No. 3. Add 5 mL of distilled water to each test tube and swirl to dissolve the crystals. Add 5 drops of 1% $FeCl_3$ to each test tube. Compare and record your observations.

2) Determining the melting point

Place a few crystals of your product which is completely dried on a watch glass and grind into powder. Prepare 2 pieces of capillary melting tubes with your sample and determine the melting point by a Thiele tube apparatus or a melting point instrument (see Experiment 13). The melting point of purified acetyl salicylic acid is 135～136 ℃[3].

Scheme 2

1. Preparation

Prepare a water bath. Heat it to 80～90 ℃ for use[1].

Weigh 6.3 g (0.045 mol) of salicylic acid and place it in a dry 50 mL Erlenmeyer flask. Use this quantity of salicylic acid to calculate the theoretical or expected yield of aspirin. Carefully add 12 mL (0.12 mol) of acetic anhydride to the flask, and add 8～10 drops of concentrated sulfuric acid.

Mix the reagents and then place the flask in the water bath. Heat the solid to completely dissolve and react for 20 min, occasionally stirring.

Remove the Erlenmeyer flask from the bath, then quickly pour the solution into a 100 mL beaker containing 50 mL of water, mix thoroughly, and place the beaker in an ice bath. The water destroys unreacted acetic anhydride and will cause the insoluble aspirin to precipitate from solution.

Collect the crystals by filtering under suction with a Büchner funnel. Wash the crystals twice with 5 mL of cold water. Carry out a complete suction until the crystals are as dry as possible. Weigh the crude aspirin.

2. Purification by recrystallization

Dissolve 5～6 g of your crude product in about 10 mL of ethanol in a 125 mL Erlenmeyer flask, warming the alcohol in a hot water bath to effect dissolution. If you obtain less than 4 g of crude product, use less alcohol in proportion.

Add 50 mL of warm water(about 50 ℃)to the alcohol solution. If there are some solids appear, heat the flask in water bath until they dissolve.

Then cool down the flask without disturbance. When crystals start to form, cool the flask by surrounding it with ice water. The crystallization process will then go to completion.

Collect the crystals by vacuum filtration. Allow the crystals to dry as possible. Weigh the product and calculate the percentage yield. Save your sample of aspirin for a melting point determination and further analysis.

3. Analysis of purity

Follow the same steps as described in Scheme 1.

Ⅴ Notes and Instructions

【Notes】

[1] It is very important to control the reaction temperature of acetylation in the range of 80～90 ℃. Some side reactions, such as side esterification or polymerization, may easily occur if the temperature is higher than 90 ℃.

[2] To avoid the hydrolysis of acetic anhydride, the graduated cylinder used for transferring and the Erlenmeyer flask used for reaction should be highly dry.

[3] Acetyl salicylic acid can decompose easily at 126～135 ℃. So the oil bath for the determination of melting point should be heated to 120 ℃ before the sample tube is inserted in order to avoid decomposing under strong heating for a long time.

【Requirements for preview】

(1) Review the acetylation reaction and learn about those acylating reagents commonly used for the synthesis.

(2) Get to know the structure of aspirin and the other synthetic methods which are different from what described in this experiment.

【Experimental precautions】

(1) Concentrated sulfuric acid is severely corrosive to eyes, skin and body tissue, so it should be handled with great care in the hood.

(2) Acetic anhydride is a strong corrosive agent and a lachrymator. Measure acetic anhydride in the hood as it is very irritating to breathe.

Ⅵ Post-lab Questions

(1) In the presence of concentrated sulfuric acid, what products can be obtained when salicylic acid is treated with ethanol? Write out the reaction equation.

(2) What is the principle of recrystallization by ethanol and water solvents?

Ⅶ Verbs

aspirin 阿司匹林；

acylating reagent 酰化试剂；

acetyl salicylic acid 乙酰水杨酸；

acetylation reaction 乙酰化反应；

acetic anhydride 乙酸酐；　　polymerization 聚合
esterification 酯化；

实验二十一　甲基橙的制备

一、实验目的

(1) 掌握用重氮盐的偶联反应来制备甲基橙的原理和方法。
(2) 学习用冰水浴控制反应温度。
(3) 熟悉固体反应产物的分离提纯方法，巩固抽滤、重结晶等操作。

二、实验原理

甲基橙是一种酸碱指示剂，它是由对氨基苯磺酸重氮盐与N,N-二甲基苯胺的乙酸盐，在弱酸性介质中偶合得到的。偶合首先得到的是嫩红色的酸式甲基橙，称为酸性黄。在碱中酸性黄转变为橙黄色的钠盐，即甲基橙。

分步反应式：

$$H_2N-C_6H_4-SO_3H + NaOH \longrightarrow H_2N-C_6H_4-SO_3Na + H_2O$$

$$H_2N-C_6H_4-SO_3Na \xrightarrow[0\sim5\ ^\circ C]{NaNO_2+HCl} [HO_3S-C_6H_4-N\equiv N]^+Cl^-$$

$$\xrightarrow[HAc]{C_6H_5N(CH_3)_2} [HO_3S-C_6H_4-N=N-C_6H_4-NH(CH_3)_2]^+Cl^-$$

$$\xrightarrow{NaOH} NaO_3S-C_6H_4-N=N-C_6H_4-N(CH_3)_2 + NaOAc + H_2O$$

本实验主要运用了芳香伯胺的重氮化反应及重氮盐的偶联反应。由于原料对氨基苯磺酸本身能生成内盐而不溶于无机酸，故采用倒重氮化法，即先将对氨基苯磺酸溶于氢氧化钠溶液，再加需要量的亚硝酸钠，然后加入稀盐酸。或者直接采用一锅法，在乙醇体系中利用对氨基苯磺酸自身的酸性使亚硝酸钠转变为亚硝酸，重氮化和偶联反应在一锅内完成。

三、仪器与试剂

【仪器】烧杯，温度计，布氏漏斗，抽滤瓶，水泵，量筒，试管，注射器。

【试剂】对氨基苯磺酸，5% NaOH 溶液，亚硝酸钠，浓盐酸，N,N-二甲基苯胺，冰乙酸，95%乙醇，乙醚。

【物理常数】

化合物	相对分子质量	密度/(g·cm^{-3})	熔点/℃	沸点/℃	溶解度/[g·(100 mL)$^{-1}$]		
					水	乙醇	乙醚
甲基橙	327.33	1.28	>300	—	0.2(易溶于热水)	不溶	不溶
对氨基苯磺酸	173.20	1.5	280	—	1.0	不溶	不溶
亚硝酸钠	69.00	2.2	320	—	81.0	不溶	不溶
N,N-二甲基苯胺	121.19	0.96	2.5	193	不溶	易溶	易溶

续表

化合物	相对分子质量	密度/(g·cm^{-3})	熔点/℃	沸点/℃	溶解度/[g·(100 mL)$^{-1}$]		
					水	乙醇	乙醚
冰乙酸	60.05	1.05	16.6	117.9	混溶	混溶	易溶
乙醚	74.12	0.71	−116.3	34.6	6.9	混溶	—

【试剂用量】

试剂	用量(方法一)	用量(方法二)
对氨基苯磺酸	2.1 g(0.012 mol)	250 mg(1.45 mmol)
亚硝酸钠	0.8 g(0.012 mol)	20%溶液 0.5 mL
浓盐酸	3 mL(0.036 mol)	—
N,N-二甲基苯胺	1.2 g(0.010 mol,1.3 mL)	125 mg(1.1 mmol)
冰乙酸	1 mL(0.017 mol)	—
95%乙醇	5 mL	2 mL
乙醚	5 mL	—
氢氧化钠	5%溶液 40 mL	0.1 g(2.5 mmol)

四、实验步骤

方法一　二步法

1. 重氮盐的制备

在 100 mL 烧杯中放置 10 mL 5% NaOH 溶液及 2.1 g(0.012 mol)对氨基苯磺酸晶体，温热使之溶解，然后冷却至室温。另溶解 0.8 g(0.012 mol)$NaNO_2$于 6 mL 水中，加入上述烧杯内，用冰盐浴冷至 0～5 ℃。在不断搅拌下，将 3 mL 浓盐酸与 10 mL 水配成的溶液缓缓滴加到上述混合液中，并控制体系温度在 5 ℃ 以下，滴加完后，在冰盐浴中放置 15 min，以保证反应完全[1]。

2. 偶联反应

在试管内混合 1.2 g(0.010 mol,1.3 mL)N,N-二甲基苯胺和 1 mL(0.017 mol)冰乙酸，在不断搅拌下，将此溶液慢慢加到上述冷却的重氮盐溶液中，加完后，继续搅拌 10 min，此时为红色液体，然后慢慢加入 25 mL 5% NaOH 溶液，直至反应物变为橙色，这时反应液呈碱性(用 pH 试纸检测，如果呈酸性可继续加入 5% NaOH 溶液调至碱性)，粗制的甲基橙呈细粒状晶体析出[2]。

3. 重结晶纯化

将反应物在沸水浴上加热 5 min，冷却至室温后，再在冰水浴中冷却，使橙黄色、呈鳞片状的甲基橙晶体析出完全。抽滤收集晶体，分别用 5 mL 冰水、5 mL 冰乙醇和 5 mL 冰乙醚依次洗涤，抽滤，烘干，称量并计算产率。

4. 性质检测

溶解少许甲基橙于水中，先滴加盐酸，再滴加 NaOH 溶液中和，观察颜色变化。

方法二　一锅法

1. 产品的合成

称取无水对氨基苯磺酸 250 mg(1.45 mmol)、N,N-二甲基苯胺 125 mg(1.1 mmol),置于 5 mL 烧杯中,再加入 2 mL 95%乙醇,用玻璃棒搅拌,在不断搅拌下,用注射器慢慢滴加 0.5 mL 20%亚硝酸钠水溶液,用水浴控制反应温度使之不超过 25 ℃。滴加完毕,继续搅拌 5 min 后,置于冰浴中冷至产品完全析出,减压抽滤,即得橙黄色、颗粒状甲基橙粗品。

2. 重结晶纯化

将产品用溶有 0.1 g NaOH 的水溶液 3～4 mL 重结晶[3],不够完全溶解时可再小心滴加少许双蒸水(每克粗产物约需 9 mL)。产物过滤干燥后称重,计算产率(约 60%)。

3. 性质检测

溶解少许甲基橙于水中,先滴加盐酸,再滴加 NaOH 溶液中和,观察颜色变化。

五、注解和实验指导

【注解】

[1] 在此时往往析出对氨基苯磺酸的重氮盐,这是因为重氮盐在水中可以电离,形成中性内盐,在低温时难溶于水而形成细小晶体析出。

[2] 湿的甲基橙在空气中受光的照射后,颜色很快变深,所以一般得到紫红色粗产物。

[3] 甲基橙在水中溶解度较大,故重结晶时不宜加过多的水。

【预习要求】

(1) 学习重氮化反应和偶联反应的机理。

(2) 理解反应温度对重氮化和偶联反应的影响。

(3) 了解甲基橙作为酸碱指示剂的应用。

【操作注意事项】

(1) 对氨基苯磺酸和 N,N-二甲基苯胺对皮肤有刺激作用,使用时要小心。

(2) 重氮化反应过程中,反应温度若高于 25 ℃,则生成的重氮盐易水解成酚,会降低产率,故用冰水浴控制反应温度非常重要。

(3) 重结晶操作应迅速,否则由于产物呈碱性,在温度高时易使产物变质,颜色变深。用乙醇、乙醚洗涤的目的是使其迅速干燥。

六、思考题

(1) 什么是重氮化反应?什么是偶联反应?简述以上两个反应的条件。

(2) 在一锅法制备甲基橙时,利用对氨基苯磺酸自身的酸性完成重氮化反应,请写出相关的反应式。

(3) 甲基橙在酸性、碱性介质中颜色和结构如何变化?

Experiment 21　Synthesis of Methyl Orange

Ⅰ Objectives

(1) To learn the method of preparing methyl orange by the diazonium coupling reaction.

(2) To learn how to control reaction temperature with water bath and ice bath.

(3) To be familiar with the fundamental operations of vacuum filtration and recrystallization.

Ⅱ Principle

Methyl orange is a pH indicator and it is very often used in titrations due to its clear color change along with the variation of pH value. In an acid it is reddish and in alkali it is yellow. In this experiment, methyl orange is synthesized by a diazonium coupling reaction of diazotized benzenesulfonic acid and N,N-dimethylaniline. The first product obtained from the coupling is a bright red acid, called helianthin. In base, helianthin is converted to the orange sodium salt, called methyl orange. The main reactions are as follows:

$$H_2N-C_6H_4-SO_3H + NaOH \longrightarrow H_2N-C_6H_4-SO_3Na + H_2O$$

$$H_2N-C_6H_4-SO_3Na \xrightarrow[0\sim5\ ^\circ C]{NaNO_2+HCl} [HO_3S-C_6H_4-N\equiv N]^+Cl^-$$

$$\xrightarrow[HAc]{C_6H_5N(CH_3)_2} [HO_3S-C_6H_4-N=N-C_6H_4-NH(CH_3)_2]^+Cl^-$$

$$\xrightarrow{NaOH} NaO_3S-C_6H_4-N=N-C_6H_4-N(CH_3)_2 + NaAc + H_2O$$

The diazotization and coupling reaction can undergo in one-pot without being isolated step by step. *p*-Aminobenzene sulfonic acid can form ethanol soluble salt with N, N-dimethylaniline. And the acidity of *p*-aminobenzene sulfonic acid make itself easily being diazotized with sodium nitrite, which give a fast coupling with N,N-dimethylaniline next.

Ⅲ Apparatus and Reagents

【Apparatus】Beaker, water aspirator, Büchner funnel, filter flask, thermometer, syringe.

【Reagents】*p*-Aminobenzene sulfonic acid, 5% sodium hydroxide, sodium nitrite, concentrated hydrochloric acid, N,N-dimethylaniline, acetic acid, 95% ethanol, diethyl ether.

【Physical constants】

Compound	M_w	ρ /(g · cm^{-3})	m. p. /℃	b. p. /℃	Solubility/[g · (100 mL)$^{-1}$]		
					Water	Ethanol	Diethyl ether
Methyl orange	327.33	1.28	>300	—	0.2(vs. in hot water)	i	i
p-Aminobenzene sulfonic acid	173.20	1.5	280	—	1.0	i	i

续表

Compound	M_w	ρ /(g · cm^{-3})	m. p. /℃	b. p. /℃	Solubility/[g · (100 mL)$^{-1}$]		
					Water	Ethanol	Diethyl ether
Sodium nitrite	69.00	2.2	320	—	81.0	i	i
N,N-dimethylaniline	121.19	0.96	2.5	193	i	vs	vs
Acetic acid	60.05	1.05	16.6	117.9	∞	∞	vs
Diethyl ether	74.12	0.71	−116.3	34.6	6.9	∞	—

Note:i,insoluble;vs,very soluble;∞,miscible.

【Reagents dosage】

Reagent	Dosage(Scheme 1)	Dosage(Scheme 2)
p-Aminobenzene sulfonic acid	2.1 g(0.012 mol)	250 mg(1.45 mmol)
Sodium nitrite	0.8 g(0.012 mol)	20% solution 0.5 mL
Concentrated hydrochloric acid	3 mL(0.036 mol)	—
N,N-dimethylaniline	1.2 g(0.010 mol,1.3 mL)	125 mg(1.1 mmol)
Acetic acid	1 mL(0.017 mol)	—
95% Ethanol	5 mL	2 mL
Diethyl ether	5 mL	—
Sodium hydroxide	5% solution 40 mL	0.1 g(2.5 mmol)

Ⅳ Procedures

Scheme 1 Two-step Method

1. Diazotization reaction

(1) In a 50 mL beaker,add 2.1 g(0.012 mol)of *p*-aminobenzene sulfonic acid and 10 mL of 5% sodium hydroxide aqueous solution. Heat the solution in a hot water bath until the solids completely dissolved. Then cool the solution to room temperature,add 0.8 g(0.012 mol)of sodium nitrite with 6 mL of water into the solution.

(2) Cool the resulting mixture in an ice bath,and add mixture of 3 mL concentrated HCl and 10 mL of water dropwise under swirling. Keep this solution cold (0～5 ℃) in the ice bath all the time. The solution should turn red-orange at this stage. Allow the resulting solution to stand in the ice bath for 15 min[1]. It now contains the diazonium salts,which will decompose if the solution becomes warm. The salts are partially soluble in the aqueous solution and will precipitate.

2. Coupling reaction

Mix 1.2 g of N,N-dimethylaniline(0.010 mol,1.3 mL)and 1 mL(0.017 mol)of acetic acid in a 10 mL beaker,then slowly add,with constant stirring,this solution to the diazonium salt suspension in the beaker. As the dull,red-purple solid start to appear,stir continuously for 10 min at 20～25 ℃ to ensure a complete reaction. A stiff paste should result in 5～10 min and 25 mL of 5% sodium hydroxide aqueous solution is then slowly added until the solution turn basic(using the pH indicating paper to detect). At the end of the coupling reaction a yellow-orange color should be observed [2].

3. Isolation of methyl orange

(1) To recrystallize the product, heat the reaction mixture in a boiling water bath for about 5 min. The solution should be translucent(though colored) under heating. Gently stir the mixture periodically to prevent bumping.

(2) Allow the mixture to cool slowly to room temperature to allow crystallization and then place the beaker in an ice bath to get it as cold as possible. Do not stir or shake the solution when it is cooling. The crystals will have a higher purity if they form slowly in a motionless process.

(3) Filter the resulting gold-orange plates of methyl orange by suction. Wash the crystals with 5 mL of cold distilled water, 5 mL of cold ethanol, and 5 mL of cold diethyl ether in turn.

(4) Transfer your product to a watch glass to air dry. Weigh the product and calculate the percentage yield.

4. Color change reaction of product at different pH

Dissolve a small amount of methyl orange crystals in distilled water in a test tube. Firstly drop dilute HCl solution then drop some dilute NaOH solution to neutralize it. Watch the color change during the process.

Scheme 2 One-pot Method

1. Preparation

In a 5 mL beaker, add 0.25 g(1.45 mmol) of *p*-aminobenzene sulfonic acid, 125 mg(1.1 mmol) of N,N-dimethylaniline and 2 mL of 95% ethanol. With constant stirring, 0.5 mL of 20% sodium nitrite solution is then slowly added through a syringe. Keep this solution cold (below 25 ℃) all the time. Continue stirring for 5 min after adding. Cool the beaker in an ice bath to get it as cold as possible to allow the orange-reddish precipitates to form completely. Filter the resulting crude product of methyl orange by a vacuum filtration.

2. Recrystallization

Dissolve your crude product in 3～4 mL of boiling water with 0.1 g of NaOH[3]. Then stand it for cooling in the air. When crystals start to form, cool the beaker by surrounding it with ice water for complete recrystallization. Filter and dry the product. Calculate the yield percentage(about 60%).

3. Color change reaction of product at different pH

Follow the same operations as described in Scheme 1.

Ⅴ Notes and Instructions

【Notes】

[1] The diazonium salt exists as a neutral inner salt which is partially soluble in the aqueous solution and may precipitate.

[2] The color of wet methyl orange is easily deepened in the air under the irradiation of light. The solid always appear to be dull and in red-purple color at this stage.

[3] Methyl orange has good water solubility at room temperature, so be careful not to add too much water during recrystallization. But if the crystals cannot dissolve completely under heating, add a little more water gradually as necessary(About 9 mL of water is needed for 1 g of dried crude products).

【Requirements for preview】

(1) Get to know the mechanisms of the diazotization and coupling reaction.

(2) Try to understand the influence of temperature to the diazotization and coupling reaction.

(3) Get to learn about the applications of methyl orange as a pH indicator.

【Experimental precautions】

(1) *p*-Aminobenzene sulfonic acid and N, N-dimethylaniline both are toxic, and N, N-dimethylaniline is readily absorbed through the skin! Handle only with gloves!

(2) The diazonium salts readily decompose even at room temperature, therefore the reaction is normally carried out in ice water bath. It is very important to control the temperature of reaction mixture for diazotization and coupling below 25 ℃.

(3) The operation of recrystallization should be quick to avoid the color change of the product. The purpose of washing the crystals with cold ethanol and cold diethyl ether is to dry the methyl orange rapidly.

Ⅵ Post-lab Questions

(1) What is a diazotization reaction? What is a coupling reaction?

(2) One-pot synthesis of methyl orange does not need extra acids for the diazotization because of the acidic property of *p*-aminobenzene sulfonic acid. Please write out the related diazotization reaction.

(3) When drop methyl orange indicator into the alkaline or acidic solution respectively, how do the structure and color of methyl orange change?

Ⅶ Verbs

diazonium salt 重氮盐;
diazotization reaction 重氮化反应;
coupling reaction 偶联反应;
one-pot method 一锅法;
helianthin 酸性黄

附　　录

附录 A　部分元素的相对原子质量
(Atomic Mass Values for Selected Elements)

元素符号(Element)	名称(Name)	相对原子质量(Atomic Mass)	元素符号(Element)	名称(Name)	相对原子质量(Atomic Mass)
H	氢(Hydrogen)	1.008	Li	锂(Lithium)	6.941
B	硼(Boron)	10.81	Na	钠(Sodium)	22.99
C	碳(Carbon)	12.01	K	钾(Potassium)	39.10
N	氮(Nitrogen)	14.01	Mg	镁(Magnesium)	24.31
O	氧(Oxygen)	16.00	Ca	钙(Calcium)	40.08
F	氟(Fluorine)	19.00	Al	铝(Aluminum)	26.98
Cl	氯(Chlorine)	35.45	Si	硅(Silicon)	28.09
Br	溴(Bromine)	79.90	P	磷(Phosphorus)	30.97
I	碘(Iodine)	126.9	S	硫(Sulfur)	32.07

附录 B 常见共沸混合物(Common Azeotropes)

附录 B-1 常见有机物与水的二元共沸混合物(Solvent-water Azeotropes)

溶剂 (Solvent)	沸点 (b. p.) /℃	共沸点 (Azeotropic Point)/℃	含水量 (Water Content)/(%)	溶剂 (Solvent)	沸点 (b. p.) /℃	共沸点 (Azeotropic Point)/℃	含水量 (Water Content)/(%)
氯仿 (Chloroform)	61.2	56.1	2.5	乙酸乙酯 (Ethyl acetate)	77.1	70.4	6.1
四氯化碳 (Carbon tetrachloride)	77.0	66.0	4.0	正丙醇 (1-Propanol)	97.2	87.7	28.8
苯 (Benzene)	80.4	69.2	8.8	异丙醇 (2-Propanol)	82.4	80.4	12.1
甲苯 (Toluene)	110.5	84.1	13.5	正丁醇 (1-Butanol)	117.7	92.2	37.5
二甲苯 (Xylenes)	140.0	92.0	35.0	异丁醇 (2-Butanol)	108.4	89.9	88.2
乙腈 (Acetonitrile)	82.0	76.0	16.0	正戊醇 (1-Pentanol)	138.3	95.4	44.7
乙醇 (Ethanol)	78.4	78.1	4.4	异戊醇 (2-Pentanol)	131.0	95.1	49.6
吡啶 (Pyridine)	115.5	92.5	40.6	氯乙醇 (2-Chloro-1-ethanol)	129.0	97.8	59.0

附录 B-2 常见有机溶剂的共沸混合物(Solvent-solvent Azeotropes)

共沸物 (Azeotrope)	组分的沸点 (b. p. of Solvents)/℃	共沸物的组成(质量分数) (Azeotropic Composition)/(%)	共沸点 (Azeotropic Point)/℃
乙醇-乙酸乙酯 (CH_3CH_2OH-$CH_3COOC_2H_5$)	78;78	3∶7	72
乙醇-苯 (CH_3CH_2OH-C_6H_6)	78;80	3∶7	68
乙醇-氯仿 (CH_3CH_2OH-$CHCl_3$)	78;61	1∶9	59
乙醇-四氯化碳 (CH_3CH_2OH-CCl_4)	78;77	2∶8	65
乙酸乙酯-四氯化碳 ($CH_3COOC_2H_5$-CCl_4)	78;77	4∶6	75
甲醇-苯 (CH_3OH-C_6H_6)	65;80	4∶6	48
乙醇-苯-水 (CH_3CH_2OH-C_6H_6-H_2O)	78;80;100	2∶7∶1	65

参考文献

[1] 周井炎. 基础化学实验(下册)[M]. 2 版. 武汉:华中科技大学出版社,2008.

[2] 陈东红. 有机化学实验[M]. 上海:华东理工大学出版社,2009.

[3] Bettelheim and Landesberg. Laboratory Experiments for General, Organic, and Biochemistry[M]. 4th Edition. New York:Harcourt,Inc,2000.

[4] Louis F Fieser, Kenneth L Williamson. Organic Experiments [M]. 7th Edition. Lexington,Massachusetts Toronto:D. C. Heath and Company,2010.

[5] Donald L Pavia,Gary M Lampman,George S Kriz,et al. Introduction to Organic Laboratory Techniques ,A Microscale Approach[M]. 4th Edition. International Edition. Brooks:Thomson Learning,2005.

[6] (德)Klaus Schwetlick 著. 有机合成实验室手册[M]. 22 版. 万均,等译. 北京:化学工业出版社,2012.

[7] John C Gilbert,Stephen F Martin. Experimental Organic Chemistry[M]. 2nd Edition. Pacific Grove,CA:Brooks Cole,1998.

[8] 陆涛,陈继俊. 有机化学实验与指导(双语)[M]. 2 版. 北京:中国医药科技出版社,2006.

[9] 薛思佳,季萍,Larry Olson. 有机化学实验(英-汉双语版)[M]. 2 版. 北京:科学出版社,2007.

[10] 李英俊,孙淑琴. 半微量有机化学实验(中英文对照)[M]. 2 版. 北京:化学工业出版社,2009.

[11] 高占先. 有机化学实验[M]. 北京:高等教育出版社,2004.

[12] 罗一鸣,唐瑞仁. 有机化学实验与指导[M]. 长沙:中南大学出版社,2012.

[13] 关鲁雄. 化学基本操作与物质制备实验[M]. 长沙:中南大学出版社,2002.